FIRE PROTECTION
SYSTEMS AND RESPONSE

T0239872

FIRE PROTECTION
SYSTEMS AND RESPONSE

ROBERT BURKE

CRC Press
Taylor & Francis Group
Boca Raton   London   New York

CRC Press is an imprint of the
Taylor & Francis Group, an informa business

CRC Press
Taylor & Francis Group
6000 Broken Sound Parkway NW, Suite 300
Boca Raton, FL 33487-2742

First issued in paperback 2020

© 2008 by Taylor & Francis Group, LLC
CRC Press is an imprint of Taylor & Francis Group, an Informa business

No claim to original U.S. Government works

ISBN-13: 978-0-367-57764-3 (pbk)
ISBN-13: 978-1-56670-622-3 (hbk)

This book contains information obtained from authentic and highly regarded sources. Reprinted material is quoted with permission, and sources are indicated. A wide variety of references are listed. Reasonable efforts have been made to publish reliable data and information, but the author and the publisher cannot assume responsibility for the validity of all materials or for the consequences of their use.

No part of this book may be reprinted, reproduced, transmitted, or utilized in any form by any electronic, mechanical, or other means, now known or hereafter invented, including photocopying, microfilming, and recording, or in any information storage or retrieval system, without written permission from the publishers.

For permission to photocopy or use material electronically from this work, please access www.copyright.com (http://www.copyright.com/) or contact the Copyright Clearance Center, Inc. (CCC) 222 Rosewood Drive, Danvers, MA 01923, 978-750-8400. CCC is a not-for-profit organization that provides licenses and registration for a variety of users. For organizations that have been granted a photocopy license by the CCC, a separate system of payment has been arranged.

Trademark Notice: Product or corporate names may be trademarks or registered trademarks, and are used only for identification and explanation without intent to infringe.

Library of Congress Cataloging-in-Publication Data

Burke, Robert, CFSP.
    Fire protection : systems and response / Robert Burke.
        p. cm.
    Includes bibliographical references and index.
    ISBN 978-1-56670-622-3 (alk. paper)

1. Fire prevention--Equipment and supplies. 2. Fire alarms. 3. Fire sprinklers. I. Title.
TH9245.B87 2008   628.9'2--dc22
2007022019

Visit the Taylor & Francis Web site at
http://www.taylorandfrancis.com

and the CRC Press Web site at
http://www.crcpress.com

Dedication

This book is affectionately dedicated to my fifteen grandchildren, Amanda, Josh, Anna, Kristin, Karlee, Kora, Kloe, Garret, Allison, Bethany, Justin, Zachary, Connor, Brandon, Isabelle, and one great-grandchild Arron (sixteenth grandchild and second great-grandchild on the way).

Contents

Preface

Author

Preface

Since taking my current job at the University of Maryland, Baltimore, I have had the opportunity to micromanage and micro-learn concerning issues of fire protection on our urban campus. Unlike my previous years in the fire service where inspectors and fire marshal's often times have such large jurisdictions that it is almost impossible to focus on any given subject for any length of time, here I have had the opportunity to really dig into and dissect almost all aspects of fire safety and fire protection. Part of that process has involved interface with the Baltimore City Fire Department (BCFD). I have thoroughly enjoyed working with the dedicated personnel of BCFD and working with old friends and making many new friends. One of my missions has been to make their job as easy as possible when they respond to alarm bells or other emergencies here at the university.

Over the past decade I have learned a great deal from the firefighters and I hope I have been able to teach them some things along the way as well. It was this interface and learning process with the firefighters of BCFD that led me to the idea of sharing some of what I have learned with other firefighters around the country. With this book I have tried to present information about subjects that I feel firefighters need to know when responding to buildings where fire alarms have been activated or where other emergencies are occurring.

It is hoped that after reading this book firefighters will have a better understanding of the workings of building fire alarm, fire protection, and other building systems. This increased level of knowledge will help firefighters to better navigate and conduct operations more efficiently and safely in all types of buildings.

Robert Burke, B.S., CFPS
Fire Marshal

The Author

Robert A. Burke, born in Nebraska, earned an A.A. in fire protection technology from Catonsville Community College and a B.S. in fire science from the University of Maryland. He has also completed graduate work at the University of Baltimore in public administration.

Mr. Burke has over twenty-eight years' experience in emergency services as a career and volunteer firefighter and has served as an assistant fire chief for the Verdigris Fire Protection District in Claremore, Oklahoma; deputy state fire marshal in the state of Nebraska; a private fire protection and hazardous materials consultant; an exercise and training officer for the Chemical Stockpile Emergency Preparedness Program (CSEPP) for the Maryland Emergency Management Agency; and is currently the fire marshal for the University of Maryland, Baltimore. Mr. Burke serves on the NFPA 45 Standard for Laboratories Using Chemicals Committee as the enforcement representative. He has served on several volunteer fire companies including West Dundee, Illinois; Carpentersville, Illinois; Sierra Volunteer Fire Company in Roswell, New Mexico; Ord, Nebraska; and Earleigh Heights Volunteer Fire Company in Severna Park, Maryland, which is a part of the Anne Arundel County fire department. He is an adjunct instructor at the National Fire Academy in Emmitsburg, Maryland, and the Community College of Baltimore County, Catonsville Campus.

Mr. Burke is also a contributing editor for *Firehouse Magazine*, with a bi-monthly column titled "Hazmat Studies" and he has also had numerous articles published in *Firehouse, Fire Chief*, and *Fire Engineering* magazines. He also writes bimonthly columns for the Firehouse.com website titled "The Street Chemist" and "Hazmat Team Spotlight." Mr. Burke has developed several CD-ROM–based training programs, including the *Emergency Response Guide Book, Hazardous Materials and Terrorism Awareness for Dispatchers and 911 Operators, Hazardous Materials and Terrorism Awareness for Law Enforcement, Chemistry of Hazardous Materials Course*, and *Chemistry of Hazardous Materials Refresher*. He has also published two books titled *Hazardous Materials Chemistry for Emergency Responders* and *Counter-Terrorism for Emergency Responders*. Both books are in their second editions. He can be reached on the Internet at robert.burke@att.net

1 Introduction

Firefighters across the United States and through out the world respond to emergencies each day in assembly, residential, commercial, business, industrial, manufacturing, and institutional buildings. These responses range from fire alarm activations to "elevator rescues" to full-blown fires and medical emergencies (figure 1.1). During these responses, personnel will be called upon to interface with a wide array of building systems including fire alarms, sprinkler systems, fire pumps, HVAC systems, elevators, emergency lighting, generators, and others. To operate effectively within these types of environments, firefighters should have a basic knowledge of the inner workings of life safety/fire protection systems and other equipment in buildings. Information found in this book is intended to provide that basic level of knowledge to emergency response personnel. It is hoped that information presented in this book will streamline fire service interaction with building features and fire protection systems. Information here may also assist designers of buildings and fire protection systems to better understand the needs of the fire service when they are called upon to operate in or near the built environment. To put this another way, architects and engineers create workplaces for firefighters. They therefore need to know what type of workplace environment is necessary for firefighters to operate effectively. Designs can be tailored to better meet operational needs of firefighters, thereby reducing the time it takes for them to mitigate an incident. One of the additional goals of this book is to reduce the deaths and injuries to responding and operating fire service personnel. When an incident can be mitigated faster, there is less time for the hazardous situation to grow in proportion. With less potential exposure, occupants of buildings and firefighters will be afforded greater protection from fire incidents.

Building codes and design standards used to construct buildings and install fire protection systems are usually well understood by designers (figure 1.2). However, many portions of these codes and standards allow design variations or contain only general performance language. The resulting flexibility permits the selection of different design options. Some options may facilitate fire service operations within a building better than others. The particular needs and requirements of the fire service are not typically well known by persons outside the fire service. It is up to fire code officials to take a more proactive role in making their needs understood during the building design process. Most times designers are happy to accommodate the needs of the fire service into building designs; they just do not always know what those needs are. The design phase of the building process is the best time to get those needs met. However, most fire service organizations do not get involved in the review process until final construction drawings are submitted for review. The application of fire protection features in buildings is similar to the consideration of the needs and comfort of building occupants when arranging a building's layout and systems. For instance, a fire code may require the installation of a fire department connection

FIGURE 1.1 Anne Arundel County Maryland Engine 26 and Truck 26 respond to alarm bells sounding in an apartment complex.

for a sprinkler system or an annunciator for a fire alarm system. However, there may be little or no guidance as to the location, position, features, or marking of such devices. Oftentimes they are hidden for aesthetic reasons. Input should always be sought and provided from local code officials and the fire service organization, one of the "clients" in this case.

Firefighters make thousands of responses each year to activated fire alarms throughout their communities. Firefighters sometimes do not fully understand the fire alarm and fire protection systems within buildings they respond to. The vast majority of fire alarm activation responses turn out to be false. False alarms keep

FIGURE 1.2 The particular needs and requirements of the fire service are not typically well known by persons outside the fire service. It is up to fire code officials to take a more proactive role in making their needs understood during the building design process.

firefighters tied up when they could be responding to real emergencies. False alarms within buildings result from numerous and varied causes, many of which are preventable. Generally, the design, installation, maintenance, and inspection issues concerning building fire alarm, fire protection, and other systems fall within the jurisdiction of the fire prevention bureau. However, "tailboard" firefighters need to take a more proactive role in the whole fire prevention and inspection process. They can also play an important role in the prevention of false alarms resulting from fire alarm and sprinkler system activations. There are those firefighters who do not believe that fire prevention or code enforcement inspections are part of their job. Firefighters nowadays have many diverse responsibilities beyond fires, including various types of rescue, EMS, hazardous materials, and, more recently, terrorism or weapons of mass destruction (WMD). Fire prevention activities, while not always popular, should be the responsibility of all firefighters. Firefighters responding to fire alarms and other emergencies in buildings are the first emergency personnel to interact with the building users/owners (figure 1.3). This encounter provides a great opportunity to make recommendations concerning false alarm prevention and other fire safety issues that may be totally unrelated to the reasons that bring them to the building. It only takes an extra few minutes to go the extra mile and make fire prevention a higher priority in the minds of all firefighters. Now, I can already hear all of the dissension noises in the background: "What, prevent false alarms? Prevent fires? My God, that's job security—if we cut down fire responses we will all lose our jobs!" Rest easy, there will still be plenty of other work for everyone, and maybe we can save some lives and property along the way. There are ways to protect life and property other than just through suppression. To quote the late fire service legend Frank Brannigan, "No firefighters ever died in a fire that was prevented." To take that one step further, no civilians ever died in a fire that was prevented, and no property was ever damaged

FIGURE 1.3 Firefighters responding to fire alarms and other emergencies in buildings are the first emergency personnel to interact with the building users/owners.

either! According to the National Fire Protection Association (NFPA), one of the leading causes of firefighter death and injury is responding to and from alarms. If we can cut down on unnecessary alarms or prevent fires from occurring in the first place, perhaps we can cut down on firefighter deaths and injuries as well. If we become more proactive as firefighters with fire prevention activities, perhaps we will prevent some fires and save civilian lives as well. The term "firefighter" does not just mean putting the wet stuff on the red stuff. You can fight fire by preventing it from starting in the first place, which I believe is a much nobler calling, even though it may not seem as glamorous or exciting. After some thought, you may realize that it is exciting to know that you may have prevented a fire from occurring and saved lives without dramatic rescue efforts. It can be just as satisfying as saving a life in a burning building and much safer for everyone involved. We just need to undertake an attitude adjustment throughout the fire service. It is my intention in this book to provide some tools to begin that process. My desire is that firefighters become better versed on the workings of fire alarm, fire protection, and other building equipment and take advantage of contact with the public to forward the cause of fire prevention. It may only take a few minutes more.

HISTORY OF FIRE PREVENTION CODES

The United States has one of the highest fire occurrence rates of any industrialized country in the World. In fact, it has been reported that the single busiest engine company in New York City responds to more fires in a year than the entire Tokyo Fire Department, which is located in the largest city in the world. Unfortunately, we are usually reactive rather than proactive when dealing with fire prevention issues and code development (Table 1.1). For example, on February 20, 2003, a fire occurred in The Station nightclub in West Warwick, Rhode Island, that killed 100 people and injured more than 187 others (Figure 1.4). This tragedy began with the improper use of pyrotechnics inside of the nightclub. Slow response by occupants to recognize the need to evacuate contributed to the difficulty in evacuating the building quickly and safely. Sounds very much like the Beverly Hills Supper Club fire in Southgate, Kentucky, in 1977; however, what is different in this case is that most occupants could see the fire. Rapid fire and smoke spread was made possible by improper interior finish materials. The nightclub was also overcrowded. Sprinkler systems were not required in the existing building by the fire codes. After the disaster, calls went out to strengthen fire codes to require sprinklers and ban pyrotechnics in this type of occupancy. Why couldn't someone have realized that this was a potential problem and fix it before those people had to die? The 2006 edition of the NFPA's *Life Safety Code*® now requires sprinklers in nightclubs with occupancy greater than one hundred. Sprinklers have been around for over one hundred years, and their record in saving lives is well documented. Why then must we wait until ten people, twenty people, fifty people, one hundred people, or five hundred people die to require sprinklers in all buildings, existing and new? Sprinklers not only save the lives of building occupants, they also save lives of firefighters. The conditions present in the Station Nightclub that contributed to the loss of life were present in other disastrous fires in U.S. history.

TABLE 1.1

Major Loss of Life Fires in the United States

Fire	Date	Lives Lost	Code Changes
Iroquois Theater, Chicago, IL	December 30, 1903	602	Outward door swing in places of assembly
Cocoanut Grove, Boston, MA	November 28, 1942	492	Egress distribution and regulation of combustible interior finishes
Marting Arms, Richmond, IN	April 5, 1968	41	Limits on gunpowder storage in retail stales occupancies
Barnum & Bailey Circus, Hartford, CT	July 6, 1944	168	Flame-retardant treatments for tents
Great Adventure Amusement Park, Jackson Township, NJ	May 11, 1984	8	Fire and life safety improvements for "amusement buildings"
One Meridian Plaza, Philadelphia, PA	February 23, 1991	3 (Firefighters)	Improvements in fire standpipe system design
Station Nightclub, West Warwick, RI	February 19, 2003	100	Sprinklers in places of assembly

FIGURE 1.4 The burned remains of The Station nightclub are visible from the air in this Friday, February 21, 2003, file photo. Three years after the fire sparked by a rock band's pyrotechnics killed one hundred people at his club, Michael Derderian, one of the brothers who co-owned it, is set to go to trial on involuntary manslaughter charges. (AP Photo/Robert E. Klein, file)

Fires at the Iroquois Theater in Chicago during December 1903 (602 dead, many children), the Cocoanut Grove nightclub in Boston on November 28, 1942 (492 dead, 166 injured), the Beverly Hills Supper Club in Kentucky on May 28, 1977 (165 dead), and Our Lady of the Angels School in Chicago on December 1, 1958 (95 dead), just minutes before dismissal, to name a few, all had similar circumstances in terms of overcrowding, lack of sprinklers, exiting problems, and interior finish issues. Changes were made to codes then, but mandatory sprinklers were not legislated or codified at that time. If they had been, likely thousands of lives would have been saved from fires that occurred between the early 1900s and the 2003 Rhode Island nightclub fire, Mizpah Hotel fire in Reno during October 2006, and the Missouri group home fire in November of 2006.

In Chicago on December 30, 1903, at approximately 3:15 in the afternoon, 602 of the 1,900 people in the audience, including many children, died in less than fifteen minutes inside the Iroquois Theater (figure 1.5). Having opened less than a month earlier, the Iroquois was advertised as a "fireproof" building although it had no fire extinguishers or fire hoses. It was not sprinkled either, even though sprinkler technology was available at the time. There had been previous loss-of-life fires in theaters in the United States and Europe. Procedures had been developed to prevent such fires; however, they were largely ignored by the management of the Iroquois. Standard precautions that had functioned well in other localities included firemen stationed near the stage with fire extinguishers, hoses, and pikes for pulling down scenery.

FIGURE 1.5 Iroquois Theatre fire, Chicago, Illinois, December 30, 1903, killed 602 people, including many children. Widely advertised as a "fireproof building," the Iroquois Theatre had opened only a month before the fire occurred. (Chicago Daily News, 1903) (Courtesy Chicago Historical Society)

In case of fire, an asbestos or iron curtain would drop down, cutting the audience off from the stage and its burning scenery. Adequate exits and trained ushers would prevent deaths from panic.

The fire began when a hot light on the stage started a velvet curtain on fire. Theaters are equipped with asbestos fire curtains, which can be deployed to separate the stage area from the audience. There was one installed at the Iroquois, but it became stuck and failed to lower all the way. The resulting gap from the failed asbestos curtain exposed the audience to smoke and fire. An actor on stage urged the audience to remain seated, which likely prevented some deaths from panic; however, some people were found dead in their seats. Exits were equipped with iron gates, and some of them remained locked. Opening the gates that were not locked required the use of a lever that most patrons were unfamiliar with, delaying the exiting of the building. Bodies were piled ten-high in the stairwell where the exits from the balconies met the main floor exits. People jumped from above the fire to the street below and many of the first ones to jump died. Others who jumped later survived the fall when the bodies of earlier jumpers cushioned their fall. The building itself was not damaged by fire and reopened a year later as the Colonial Theater; it remained in business until it was torn down in 1925. Several fire protection features were absent in the building that could have reduced the large loss of life in the theater. These include:

- Blocked asbestos curtains
- Installed ventilators that were not in operation
- Exits not properly marked
- Exits blocked with draperies, wood, and glass doors
- No installed alarm system
- No fire protection devices such as extinguishers, hoses, or standpipes
- No automatic sprinklers in the stage area, even though it was a municipal requirement at the time

The investigation that followed led to a variety of fire safety improvements, all of which addressed the problems listed above. Following the Iroquois Theatre fire, all Chicago theaters in the city were closed down for a period of time and required to be brought up to new codes passed by the city. Although requirements for sprinklers were removed from the new code revisions, there were some improvements to the codes. Balconies were required to have distinct and separate places of exit and entrance. Aisles were required to be a minimum of thirty inches wide with corridors no less than forty-eight inches wide. No more than fourteen seats were allowed to be located in any one row between aisles. Rows have to be thirty-two inches apart back-to-back. Exit doors could not be obstructed by curtains, locked, or fastened in any way during the time the building was occupied. Exit signs had to be provided, marking the way out of the building. Scenery and props had to be treated with paint or a chemical solution to make them flameproof. Flues and smoke vents were required to be opened and closed by a closed-circuit battery and controlled by two newly situated switches, one at the electrician's station on the stage, which was required to be fireproof, and the other at the city fireman's station. As has been true throughout history, once again the real life-saving requirements for sprinklers in Chicago's theaters were left out of the requirements. In fact, at the time of the fire, sprinklers were required, and the new code revisions eliminated the sprinkler requirement. So, in effect the new fire code revisions following the Iroquois Theater fire were actually inferior to the existing code at the time of the fire.

Fire at the Rhythm Night Club in Natchez, Mississippi, on April 23, 1940, resulted in the deaths of 209 African Americans and severely injured many others. There were over 300 people in the building at the time of the fire. The nightclub was housed in a converted church and former blacksmith shop and was a single-story wood-frame building. This fire remains the second deadliest nightclub fire in the United States. The fire started around 11:30 p.m. in front of the main entrance to the building by two women carelessly discarding a match. Decorative Spanish moss draped over the rafters led to rapid flame spread quickly engulfing the structure. It is believed that highly flammable methane gas was generated by the moss, leading to the rapid fire spread. Windows were boarded up to keep people outside from viewing the activities inside the club. Most people died from smoke inhalation or from being crushed in the furious stampede of people trying to escape. The building was overcrowded at the time of the fire and had no fire alarm or sprinkler systems.

Another tragic fire occurred on Saturday, November 28, 1942, in Boston at the Cocoanut Grove nightclub (figure 1.6). There were approximately 1,000 occupants in the building at the time the fire broke out at 10:11 p.m. The occupancy limit for the building was set at 490. A busboy using a match to change a burned-out lightbulb in the basement is thought to have caused the fire. Exits at the front of the building consisted of two revolving doors. These doors quickly became blocked and were piled with body's four- to five-deep. Four hundred ninety-two people died in the fire and

FIGURE 1.6 Boston City fire and police department workers and many servicemen jam the street outside the Cocoanut Grove nightclub in Boston, Massachusetts, on November 28, 1942, helping to haul out victims of a sudden fire that swept the club before football weekend patrons had a chance to escape (AP Wire Photo).

experts believe that as many as three hundred of those people might have survived had these main exit doors been side-hinged and swung out from the building, as is the requirement in assembly occupancies today. It is ironic that the number of dead is almost exactly the number of people that were over the occupancy limit! A panic bar on another exit had been bolted or welded shut (both are listed in different accounts of the fire). Many bodies piled up by this unusable exit as well. It is believed that there was also a refrigerant leak of highly flammable methyl chloride in the building, which may have contributed to the high death toll. Freon was in short supply because of World War II, which was going on at the same time. Many of the patrons in the club were in the military. Some people dining in the restaurant died in their seats. Flammable decorations also contributed to the fast spread of the fire. As a result of the Cocoanut Grove fire, side-hinged, outward-swinging exit doors with panic bars were required on either side of revolving doors. Building occupancy capacity also was required to be posted on signs in assembly occupancies as a result of the fire. Illuminated exit signs and emergency lighting became requirements as a result of the Cocoanut Grove fire as well.

Yet another nightclub fire occurred on May 28, 1977, at the Beverly Hills Supper Club in Southgate, Kentucky, a suburb of Cincinnati, Ohio. The fire began around 9:00 p.m., and before it was over, 165 people had lost their lives. It was Memorial Day weekend and between 2,400 and 2,800 people were crowded into the club. Previous patrons had reported the Beverly Hills Supper Club was disorienting because of its many rooms and corridors. The Beverly Hills Supper Club had been remodeled in the past along with construction of several additions to the building. It should be pointed out that automatic sprinklers, fire alarm systems, and kitchen hood fire extinguishing equipment was not installed during any of the construction projects. At about 8:45 p.m. on May 28th, employees discovered a fire in the Zebra Room (figure 1.7). There appears to have been about a fifteen-minute delay in notifying the fire department. During this time, employees attempted to extinguish the fire themselves. This fire occurred thirty-five years after the Cocoanut Grove fire in Boston, but many of the contributing factors at Cocoanut Grove were also present at the Beverly Hills Supper Club. There were no fire protection systems installed, no fire safety plan, blocked exits, crowds in excess of the occupancy load, inadequate exit capacity, and combustible wall coverings. The official investigations into what caused the fire at the Beverly Hills Supper Club were inconclusive, but the factor most often cited is aluminum wiring. A federal jury in Ashland, Kentucky, agreed, ruling in July 1985 that aluminum wiring had caused the fire. According to a report from a commission appointed by the governor, there was little or no code enforcement at the supper club. Inspectors were often given free meals and drinks.

During December of 1958, Our Lady of the Angels School in Chicago caught fire, killing ninety-two students and three nuns and injuring seventy-seven others (figure 1.8). Smoke and fire cut occupants off from their exits, which consisted of open corridors and open stairwells with no barriers to stop the spread of smoke and fire. At the time of the fire there were approximately 1,600 students in kindergarten through eighth grade. While legally in compliance with the fire safety codes of the time, the school was sadly unprepared for any kind of fire. There was only one fire escape, no sprinklers, no automatic fire alarm, no smoke or heat detectors, no alarm connected to the fire

FIGURE 1.7 Southgate, Ky., May 30—Picking Through the Ruins—Firemen pick through the remains of the gutted Beverly Hills Supper Club Monday. The night club, located about five miles south of Cincinnati, was gutted by flame Saturday night, killing more than 160 persons (AP Wire Photo).

department, no fire-resistant stairwells, and no fire-rated doors from the stairwells to the second floor. Windows from the second floor were twenty-five feet above the ground. The fire started in the basement sometime between 2:00 and 2:20 p.m. in a cardboard trash barrel at the foot of the northeast stairwell. The fire burned undetected

FIGURE 1.8 During December of 1958, Our Lady of the Angels School in Chicago, Illinois, caught fire, killing ninety-two students and three nuns and injuring seventy-seven others (Courtesy OLA Website, www.olafire.com).

for an estimated fifteen to thirty minutes, gradually filling the stairwell with super hot gases and smoke. For 329 children and 5 teaching nuns, the only remaining means of escape was to jump from their second floor windows to the concrete and crushed rock twenty-five feet below or to pray for the fire department to arrive and rescue them before it was too late. Recognizing the trap they were in, some of the nuns encouraged the children to sit at their desks or gather in a semicircle and pray, and they did—until the smoke, heat, and flames forced them to the windows. But there were no firefighters to rescue them. Some began jumping; others fell or were pushed. Of those who jumped, some were killed in the fall, and scores more were injured. Many of the smaller children were trapped behind the frantic crowds at the windows, blocking any chance to escape through a window. Many of the little ones who managed to secure a spot at a window were then unable to climb over the three-foot-high windowsills or were pulled back by others frantically trying to scramble their way out. Helplessly, firefighters watched in horror as classrooms still filled with frightened children exploded in flames, instantly killing those who remained. The first fire department units had arrived within four minutes of being called, but by then the fire had burned unchecked for as long as thirty minutes and was raging out of control. Firefighters also were delayed upon arrival because they had been incorrectly dispatched to the rectory around the corner and lost valuable minutes repositioning their apparatus and hose lines after realizing the true location of the fire.

Although the cause has never been officially determined, all indications point to arson. A boy (age ten at the time, a fifth grader in room 206) later confessed to setting the blaze but subsequently recanted his confession. Officially, the cause of the fire remains unknown. Our Lady of the Angels School passed a fire department safety inspection only weeks before the fire, primarily because the school did not have to comply with all fire safety guidelines due to a grandfathering clause in the 1949 standards. Existing schools were not required to retrofit the fire protection and alarm systems that were required in all newly constructed schools. In the only positive outcome of the tragedy, sweeping changes in school fire safety regulations were enacted nationwide, no doubt saving countless lives in subsequent years. Some of the changes in regulations included:

- Mandatory exit drills
- Construction practices according to code
- More school inspections
- Greater emphasis on installed fire protection, alarms, and first aid firefighting equipment

As a result of the tragedy at Our Lady of the Angels in Chicago, ordinances to strengthen Chicago's fire code and new amendments to the state fire code were passed. Additionally, the NFPA estimated that hundreds of schools across the nation were safer because, according to an NFPA survey, about 68 percent of all U.S. communities inaugurated and completed fire safety projects after the Our Lady of the Angels holocaust.

As recently as October 31, 2006, a fire at the historic eighty-four-year-old Mizpah Hotel, a landmark in Reno, Nevada, claimed eleven lives, injuring thirty others leaving three in critical condition. It was the largest loss-of-life fire in Reno history. The building housed mainly borders, people who lived there from week to week. There were approximately eighty people in the hotel when the fire broke out. The hotel had smoke detectors but no sprinklers. Once again, another older building that was grandfathered in the fire codes and not required to install sprinklers that could have saved many if not all of the eleven people who died. How much value do you place on a person's life? Were eleven people worth the savings that were allotted the owners by not having to install sprinklers? Fortunately, no firefighters were killed or seriously injured fighting the fire. A forty-seven-year-old female resident of the hotel was arrested for setting the fire, which she started by setting fire to a mattress.

Monday, November 28, 2006, another multiple loss-of-life fire occurred in Anderson, Missouri, in a group home for the mentally ill. Ten people were killed and twenty-five others injured when the fire broke out around 1:00 a.m. The building was a total loss and was nearly fully involved in fire when firefighters arrived on scene. The facility was licensed by the Missouri Department of Mental Health to allow mentally ill residents to live at the home and receive treatment elsewhere. There had been another small fire at the home the previous Saturday. Causes of the fires are still under investigation at the time of the printing of this book. The facility was not sprinkled.

Large loss-of-life fires grab the media headlines and result in major changes in the codes, but what about the disaster of residential fires that occur each and every year? There are more people killed in residential fires each year then in all of the natural disasters combined. Maybe if they all died at once we would react and require residential sprinklers. Thirty-five hundred–plus people killed in residential fires every year is a disaster and a tragedy because most could have been saved (table 1.2). The number of deaths initially declined with the development and widespread use of smoke alarms, but in the last several years deaths are on the rise again. Quick response residential sprinkler technology has been available for many years. Most people who die in fires die in residential fires. The solution is obvious: require sprinklers in all multi-family and single-family occupancies; the resulting decline in fire deaths would be dramatic. Often when codes are changed to require sprinklers in particular occupancies, they seldom apply to existing buildings. Existing buildings are usually the ones that really need the code changes to upgrade the life safety of the occupancy. However, we are reluctant to place the burden of code changes on current building owners because of the potential cost and disruption of operations of the business. So we wait until disaster strikes and then attempt to change what we knew all along should have been changed. We have the technology to prevent such disasters; why then do we not become proactive and make sprinklers mandatory in all occupied structures, including multi-family and single-family occupancies?

In the United States we enact fire codes following major loss-of-life disasters that show us where we had a code deficiency. We still do not take advantage of all the technology available to reduce the impact of fires on life and property before they happen. Every building constructed in this county, including single-family homes, should be built with a complete automatic fire sprinkler system. Sprinkler systems have a proven safety record for saving life and property. According to the NFPA,

TABLE 1.2
Fire in Residential Structures

Year	Fires	Deaths	Injuries	Dollar Loss (Millions)
1996	428,000	4,080	19,300	4,962
1997	406,500	3,390	17,775	4,585
1998	381,500	3,250	17,175	4,391
1999	383,000	2,920	16,425	5,092
2000	379,500	3,445	17,400	5,674
2001	396,500	3,140	15,575	5,643
2002	401,000	2,695	14,050	6,055
2003	402,000	3,165	14,075	6,074
2004	402,000	3,225	14,175	5,948
2005	396,000	3,055	13,825	6,875

there has never been a multiple loss-of-life fire in a fully sprinkled building, where the sprinkler system was not damaged by an explosion or other trauma (NFPA considers a multiple loss of life to be three or more people; figure 1.9). That is an amazing record considering the frequent reports in the news media of multiple loss-of-life fires occurring throughout the United States. As I write this chapter, news has arrived indicating that two Memphis, Tennessee, firefighters have died fighting a fire in a

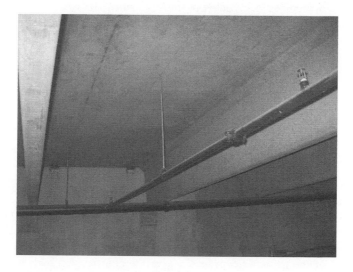

FIGURE 1.9 Sprinkler systems have a proven safety record for saving life and property. According to the National Fire Protection Association (NFPA), there has never been a multiple loss-of-life fire in a fully sprinkled building where the sprinkler system was not damaged by an explosion or other trauma.

single-story Dollar General Store. Sprinklers were not required in the building. But had they been in place, those two firefighters might still be alive today. Then there was the McDonald's restaurant fire in Houston, Texas, that killed two firefighters. McDonald's and other fast-food restaurants are not large buildings. So how much could it cost to sprinkle them? What if the building had been sprinkled? Would those firefighters be alive today? We continue to kill civilians and firefighters when sprinkler systems could have saved their lives. Sprinkler systems can also help with departmental budgets. Fully sprinkled buildings do not require the same resources in terms of personnel and equipment to fight a fire compared to an unsprinkled building. Most times firefighters just have to "mop up" after the sprinkler system has controlled or extinguished the fire. Fire companies can go back into service quicker when a sprinkler controls or extinguishes a fire. That is less wear and tear on equipment, fewer repairs, less fuel, and less wear and tear on personnel.

The leading cause of firefighter line-of-duty deaths is from heart attacks. We beat firefighters down fighting fires in all types of buildings. This stress certainly contributes to heart attacks, even if it is not the total cause. Recently it has been reported that hydrogen cyanide may be responsible for deaths and injuries from smoke inhalation because it is present in the smoke at many fires. It has also been linked to heart problems. Maybe we are finally finding the link to the annual firefighter deaths from heart attacks. Maybe if all buildings were sprinkled, firefighters would not be exposed to so much smoke and the hydrogen cyanide and other toxic gases it contains. Maybe we could finally start reducing the number of annual firefighter deaths. If all buildings were sprinkled, we would not need as many large fire engines with high-capacity pumps, aerial devices, and other suppression-related equipment. For many fire departments, medical calls make up as much as 70 percent of the call volume. First-response vehicles could be custom-made to more accurately reflect the types of alarms most departments respond to if all buildings were sprinkled. The vehicles would not be as big or expensive, reducing the costs of firefighting, which would provide relief to many departments with budget shortfalls.

So what is going on? We have identified the problem: fires occur, killing thousands of civilians and close to one hundred firefighters or more each year and causing millions of dollars in property damage. As I mentioned before, fires cause more death and injury and property damage each year than all natural disasters combined. Yet we often spend more time and money planning for and mitigating natural disasters than we do for the prevention of fires. We have technology proven to reduce both the loss of life and property damage. So, why do we not do something about implementing that technology? The answer is not a simple one. It involves issues including changing public perception, attitudes, politics, and leadership. Most civilians and even some in the fire service do not believe fire will happen to them. Fires are something we hear about in the news, but they always happen to someone else. When sprinkler systems are proposed in buildings where not required by code, they are often one of the first things to be cut from the construction budget (sometimes referred to as "value engineering"). What value is there to removing from a building a system that may save lives once the building is occupied, not to mention preservation of the building itself? With the right attitude, sprinkler systems would be untouchable by budget cuts.

Politically, mandatory sprinkler systems are a tough sell to legislators and administrators. After all, business owners are voters, and making them bring buildings up to code and install sprinkler systems may not be popular with the voters. Often architects and construction concerns lobby against mandatory sprinkler requirements because they increase the cost of a project. Unfortunately, fire service leaders in many cases are at fault. They need to advocate and justify the cost-effectiveness of mandatory sprinkler systems and make it a nonnegotiable issue. Mandatory sprinkler systems will eventually reduce operational costs of the fire department. It takes bold leadership among fire service administrators, political leaders, and legislators to enact mandatory sprinkler legislation. That leadership needs to start at the federal level at the United States Fire Administration (USFA). USFA has been moved back as a part of the Federal Emergency Management Agency (FEMA), which is a part of the Department of Homeland Security. FEMA has been spending billions of dollars each year to respond to and mitigate natural disasters, most of which cannot be prevented. It seems to me that we ought to be spending that kind of money on fire, the "forgotten disaster" in this country. In those cases where fire does occur, it can be controlled or extinguished by automatic sprinkler systems. The federal government needs to tie strings to state aid from USFA/FEMA/Homeland Security based upon enacting mandatory sprinkler legislation. This could be done much like they do for highway and other transportation funding that is tied to compliance to certain federal programs and regulations for highway safety: No mandatory sprinklers, no federal grants. In fact, if we had mandatory sprinkler legislation across the country, maybe there would not be as much need for fire grants! USFA needs to come out into the open and push for mandatory sprinkler systems instead of proposing that we just put them in kitchens as a start. Leadership also needs to come from code development organizations, including NFPA, the *International Building Code*, and others. Codes need to require mandatory sprinkler systems in all types of occupancies. In the 2006 NFPA 101 *Life Safety Code*, bars with live entertainment, dance halls, discotheques, nightclubs, and assembly occupancies with festival seating with occupancy greater than one hundred are now required to be sprinkled. This is a direct result of the Station Nightclub fire in Warwick, Rhode Island. But NFPA stops short of requiring sprinklers in all assembly occupancies. NFPA allows other types of assembly occupancies with less than three hundred people to not be sprinkled. So now we have to wait for more people to be killed to get those types of occupancies sprinkled when we know already that it should be done now. NFPA 101, the *Life Safety Code*, also now requires sprinklers in all new one- and two-family dwellings. It's about time, but what about existing residential occupancies? Some states and local jurisdictions that have adopted NFPA 101 have amended it to remove the sprinkler requirements for existing and some new one- and two-family dwellings. How can you justify that? One- and two-family dwellings are where most people die in fires. Why should sprinkler requirements be amended out? What's wrong with this picture? Some people still do not get it.

Housing fires both on and off campus at colleges and universities is another issue that has been given a lot of press over the past several years. The issue really got its start with the fire at a dormitory at Seton Hall University located in East Orange, New Jersey, on January 19, 2000. Three students were killed and fifty-eight others injured.

The fire at Seton Hall was not the worst dorm fire on a college campus or the most recent. Once again, history repeats itself because of a failure of college administrations to provide—and code officials to require—sprinkler systems in college dorms. The carnage does not end with the Seton Hall fire; nationwide, ninety-four people have been killed in campus-related fires since January 2000. How many more students have to die? Following the Seton Hall fire there was a big push to get all on-campus student housing sprinkled across the country. At the University of Maryland, Baltimore, we do not have a large number of student housing units; however, we were able to get all of the units sprinkled by continuing to press the issue over several years. Now the focus has turned to off-campus housing and Greek housing. On November 18, 2006, one student was killed and three others critically injured at a fire at the Phi Kappa Tau fraternity house at Nebraska Wesleyan University in Lincoln, Nebraska (figure 1.10). Six weeks following the fire, Nebraska Wesleyan University has begun adding sprinklers to all resident halls and fraternity houses. It's just a shame that one student had to die and three others were critically injured to get the sprinklers installed. After all, this is six years after the Seton Hall Fire. Administrators at the colleges involved are now pushing for sprinklers in all student and Greek housing units on campus. Off-campus housing is also an issue. Many of these off-campus occupancies are converted single-family or converted multi-family homes. Few if any of them are regulated in any way, and many are simply firetraps. A student from the University of Pittsburgh died from injuries he suffered in an off-campus fire on November 4, 2006. Once again there were no sprinklers present in the structure that burned. On December 19, 2006, another student died in off-campus housing, this time in an apartment building in Lincoln, Nebraska (the second student death in Lincoln in just a month). A pregnant student and her unborn child died in the fire, and her 2½-year-old daughter was injured. The mother was scheduled to have her baby the

FIGURE 1.10 Members of the Lincoln Fire Department start rolling up hoses at the scene of a fire Friday, November 18, 2006, at Phi Kappa Tau fraternity house at Nebraska Wesleyan University in Lincoln, Nebraska. The fire killed one student and left three others critically injured (AP Photo/*Lincoln Journal Star*, Ted Kirk).

next day. Just a simple room and contents fire that a single sprinkler would have likely extinguished, but there were no sprinklers installed in the apartment building.

Leadership to require the installation of sprinkler systems needs to occur at the state and local level as well. Fire marshals in every state need to take a more proactive role in selling the idea of mandatory sprinkler systems to governors, legislators, and the general public. Fire chiefs need to sell the idea of mandatory sprinklers in all buildings to the local city and county legislative bodies. It is past time to put a stop to the tragic loss of life to fire in unsprinkled buildings. Just this week I heard in the news of eleven children who died in home fires, six in Michigan and five in Mississippi. These were situations where "good sense–challenged" parents left young children unattended at home. However, if there had been sprinklers in these homes, they likely would not have died. (I am not advocating substituting sprinklers for proper supervision of children!) When will we say enough is enough and take steps to stop the carnage?

FIRE DRILLS

How many fire departments out there attend fire drills held in buildings in your communities other than schools (figure 1.11)? How many departments even know whether fire drills are being held and in what buildings? How many departments have never even given the idea of fire drills a second thought? How many departments think they are just too busy to deal with fire drills? Fire drills are required in NFPA standards and state and local law in various types of occupancies. NFPA promotes fire drills in the home during National Fire Prevention Week held each October. I know

FIGURE 1.11 How many fire departments out there attend fire drills held in buildings in your communities other than schools? How many departments even know whether fire drills are being held and in what buildings? How many departments have never even given the idea of fire drills a second thought?

that the home is where most fire deaths occur. But what about fire drills in other types of occupancies? Schools and some government buildings hold regular fire drills, which are often required by state law. But what about fire drills in the dormitory, the apartment building, the office, the factory, the assembly, or the mercantile occupancy? What about the numerous false alarms, which cause building occupants to shy away from evacuation when a drill or real alarm occurs? Would fire drills have prevented the six deaths in the Cook County, Illinois, high-rise office building fire during October 2003? There was certainly confusion about proper evacuation procedures. The building did not have sprinklers either—a thirty-five-story building with no sprinklers? Here we go again. Would sprinklers have saved the six lives? It is very likely they would have. Just how many more people have to die?

On February 5, 1999, at 1:00 a.m., a fire occurred in the Charles Towers Apartment Complex, a thirty-two-story high-rise with 250 occupants in downtown Baltimore, Maryland. This facility had been plagued with false alarms in the past, and once again there were no sprinklers. A thirty-two-story residential occupancy with no sprinklers! Enough is enough! Why was this building not retrofitted with sprinklers? When the real fire occurred many residents failed to evacuate when it was still safe to do so. Firefighters had to be lowered onto the roof of the building from a helicopter to rescue those who did not evacuate and could not get out because they waited too long. Fortunately, everyone was rescued safely and no firefighters were killed or seriously injured. No sprinklers were added after the fire. No code changes in the city to require existing apartment buildings or high-rise buildings to be sprinkled. Had there been multiple deaths perhaps sprinklers would have been called for. However, this kind of apathy can lead to disaster. Fire departments must do everything they can to prevent false alarms and make sure that fire drills are conducted at target hazards in their community. An unsprinkled thirty-two-story residential occupancy should certainly be a target hazard, especially a building with such a poor history of false alarms. Fire drills can ultimately help firefighters to evacuate buildings when a fire or other emergency occurs. We will not always be as lucky as they were at Charles Towers. Do we want to continue to play Russian roulette with fires in occupied buildings?

OCCUPANT EVACUATION ASSISTANCE

Building occupants can be trained to assist in evacuation when a fire alarm sounds or when a fire or other emergency breaks out (figure 1.12). Occupant assistants are referred to by various names, but for this book we will refer to them as "fire wardens." Fire warden systems can be developed in many types of occupancies to assure that people begin the evacuation process as soon as the fire alarm sounds (there should never be a delay in calling the fire department or beginning the evacuation of a building when the fire alarm sounds). Fire wardens should be provided training to assist them in carrying out their job in an efficient and expedient manner. When routine inspections are conducted at facilities in the community or when firefighters respond to the building during an emergency, they should quiz occupants about the existence of false alarm problems and fire drill frequency. Ask them if they have any type of fire warden system. Develop literature and training classes that can be handed off to

FIGURE 1.12 Building occupants can be trained to assist in evacuation when a fire alarm sounds or when a fire or other emergency breaks out. Occupant assistants are referred to by various names, but for this book we will refer to them as "fire wardens."

occupants to assist them in implementing fire warden programs. Training programs can be on video, PowerPoint, CD, DVD, online, or live. Firefighters should supervise drills that do occur as time permits and make suggestions on improving drill performance. Fire drill information is available from insurance companies, the NFPA, the USFA, and others. When a real fire does occur and firefighters arrive and find that the building is already evacuated, their job is just that much easier. If there is a sprinkler system, their job will also be very short! Having a system of fire wardens and occupant accountability reduces or eliminates the need for secondary searches. After all, how many firefighters have been killed or injured doing searches for people who were not even there? Once again it's time to become proactive in preventing false alarms and encouraging fire drills. Fire drills do help to save lives.

STANDARDS AND CODES OVERVIEW

References in this book to NFPA and other codes are provided for informational purposes only and should not be used as a substitute for actual code research. This chapter and many of the other chapters in this book make reference to code requirements but do not list the actual code citations. Information provided here is an interpretation of sections of the code by the author and the full text of the code should be used when evaluating any situation in which the code applies. Local and state amendments to codes may alter information provided here.

NFPA standards are sometimes looked upon for specialized technology areas such as suppression systems, fire alarms, standpipes, fire pumps, emergency lighting, generators, water supplies, and many others. One thing to remember is that all codes are minimum standards. There is nothing that says we cannot initiate stricter local standards through amendments to the implementation document, which enacts the code by the local government. Handbooks are written for some common NFPA codes, such as the *Life Safety Code*, sprinkler code, and others. These handbooks are designed to help explain the concepts behind code requirements, in some cases what the code writers were thinking and clarification of confusing issues. Codes also have explanatory information for many sections, which are located in the appendix of the code. If a particular NFPA code section has an asterisk (*) following the code section number, this means there is explanatory information located in the appendix. This information can help justify and explain code requirements to the public. NFPA and other codes are developed by technical committees composed primarily of volunteers from various organizations both public and private and are issued on a regular cycle (usually every three years). Code technical committees are comprised of experts in a particular subject area, many times including fire officials (although there are those, including this author, who believes the fire service is underrepresented on codes development technical committees). Technical committees can issue interpretations of code questions formally submitted to them for review. These interpretations can be helpful in applying the code to local or unusual situations. I was personally appointed to the NFPA 45 Committee on Laboratories Using Chemicals and am presently a member starting my second revision cycle. The *Fire Protection Handbook* published by the NFPA can also provide useful technical information on a broad variety of fire protection issues.

NFPA has a certification program available for those fire personnel wanting to become a certified fire protection specialist (CFPS) or certifications at other occupational levels including but not limited to fire inspector, plans examiner, and fire investigator. CFPS certification carries significant weight in terms of status within the fire protection community. The only thing better in terms of certification would be to become a fire protection engineer. Certified fire protection specialists are tested on their knowledge of and their ability to locate information in the *Fire Protection Handbook* in order to obtain certification. Pre-test seminars are often presented by the NFPA, state fire training organizations, and others to prepare personnel for taking the examination, which, by the way, is very grueling. Seminars are also available from the NFPA concerning specific issues of fire protection including the *National Electric Code*® (NFPA 70), *National Fire Alarm Code*® (NFPA 72), *Sprinkler Systems* (13, 13D, 13R), and others.

FIREFIGHTER KNOWLEDGE OF FIRE PROTECTION SYSTEMS

Firefighters should have a general working knowledge of fire alarm and fire protection and other building systems installed in various buildings (figure 1.13). They need to become familiar with fire alarm system components, including the annunciator panel; fire alarm panel; sprinkler system components; shutoff and control valves; smoke control and evacuation equipment; stair pressurization equipment; elevator fire service and emergency controls; emergency lighting; HVAC systems and controls; and fire command center operations in high-rise buildings. Personnel who work in the fire prevention bureau should have an in-depth knowledge of fire alarm and fire protection systems. They also need to have a good working knowledge

FIGURE 1.13 Firefighters should have a general working knowledge of fire alarm and fire protection systems installed in various buildings.

of the fire codes relevant to such systems and be able to apply those codes to shop drawings, make recommendations, and approve installations. Prevention personnel should become certified as fire inspector I, II, or III depending on their usual duties. Certification as a fire protection specialist (CFPS) will also add creditability to their prevention and code enforcement efforts. Training for the above-mentioned certifications is available from a number of sources. NFPA has recently developed a new level of certification for fire marshals. Proposed standard NFPA 1037 *Standard for Professional Qualifications for Fire Marshal* will identify the professional level of performance required for fire marshal. The standard will specify job performance requirements for the minimum qualifications for professional competence for fire marshals and equivalent positions. This standard is currently in the draft stage. Local and state training organizations, community college fire science programs, the NFPA, the Society of Fire Protection Engineers, and the National Fire Academy provide seminars and training programs for fire inspectors and fire marshals. Membership in peer group associations can be helpful in obtaining information and training for fire prevention technology issues. Many states have fire marshal, fire inspector, or code official organizations. NFPA members can join the International Association of Fire Marshals. There are plenty of resources available for firefighters wishing to gain more knowledge about building systems.

REQUIREMENTS FOR FACILITY FIRE PROTECTION

Until we have the courage to make sprinklers mandatory in all types of occupancies, sprinkler system and alarm system requirements are tied to specific types of occupancies. Occupant load as well is directly connected to occupancy type (table 1.3).

TABLE 1.3
NFPA Occupancy Classifications and Occupant Load Factors

Occupancy Classification	Ft² (Per Person)	m² (Per Person)
Assembly concentrated use (no fixed seats) < 10,000 ft²	7	0.65
Assembly concentrated use (no fixed seats) > 10,000 ft²	5	
Assembly less concentrated use (no fixed seats)	15	1.4
Assembly bench-type seating	18 linear inches	45.7 linear centimeters
Assembly fixed seating	Number of fixed seats	Number of fixed seats
Educational classrooms	20	1.9
Educational shops, laboratories, vocational rooms	50	4.6
Day care use	35	3.3
Health care use; inpatient treatment departments	240	22.3
Health care use; sleeping departments	120	11.1
Detention and correctional	120	11.1
Residential use; hotels and dormitories, apartment buildings, board and care, large	200	18.6
Industrial use	100	9.3
Business use	100	9.3
Mercantile use; sales area, street floor	30	2.8
Mercantile use; sales area, on two or more street floors	40	3.7
Mercantile use; sales area, on floor below street floor	30	2.8
Mercantile use; sales area, on floors above the street floor	60	5.6
Mercantile use; floors or portions of floors for storage, receiving, and shipping and not open to general public	300	27.9

Note: Reprinted with permission from NFPA 101®, *Life Safety Code*®, Copyright© 2006, National Fire Protection Association, Quincy, MA 02169. This reprinted material is not the complete and official position of the National Fire Protection Association on the referenced subject, which is represented only by the standard in its entirety.

For example, in business occupancies listed in the NFPA *Life Safety Code*, the occupant load is one person per one hundred square feet. In an assembly occupancy without fixed seating it is one person per fifteen square feet. Firefighters should be familiar with the occupancy classifications and occupant load requirements of the various codes. They need to be able to determine whether an occupancy is overcrowded during inspections or emergency responses. Fire prevention personnel should have occupancy load limit signs placed in assembly occupancies. This will assist suppression personnel responding to an alarm or conducting a company-level inspection in knowing whether an occupancy is overcrowded. NFPA's *Life Safety Code*, the local building codes, and within the following listed specific codes there is a great deal of difference in what protection requirements exist depending on the type of occupancy. For example, health care and educational occupancies have much more stringent code requirements for alarm and protection systems than do business occupancies. Several years ago, colleges and universities successfully lobbied the NFPA to be moved from the educational occupancy classification to the business occupancy. Business occupancy requirements are much less stringent than those for educational occupancies. Knowing the occupancy type and occupant load calculations can also give you an idea of the numbers of people who may be expected within a facility when an actual emergency occurs. Codes also address special structures such as towers, high-rise buildings, and tents. Firefighters do not need to memorize any of the codes. They just need a general knowledge of what codes apply to what situations and know where to look to answer code issues and questions. NFPA 1, the *Fire Prevention Code*, has the most common excerpts from other NFPA codes to give firefighters and code officials a condensed book to take with them on inspections to address common deficiencies.

NFPA Codes Overview

NFPA 10 *Standard for Portable Fire Extinguishers®* provides code information regarding portable fire extinguishers. This information includes selection of the proper type of extinguisher (which is related to the hazard), size and distribution of extinguishers, inspection and record-keeping, maintenance, recharging and testing, and hydrostat test frequencies. Because some fire extinguishing agents can be corrosive and most are stored in pressure containers until needed, the containers must be hydrostatically tested periodically to insure container integrity. All fire extinguishers when encountered in buildings should meet the requirements of NFPA 10. The fact that fire extinguishers are located within a building to begin with is usually a result of OSHA or individual code occupancy chapter requirements. Fire extinguishers may also be placed inside buildings by owners or occupants as a good safety practice. But, no matter how they got there to begin with, once provided, fire extinguishers must be properly tested and maintained. It is also important that occupants be trained to use portable fire extinguishers. If occupants are not trained, they should be instructed not to use the extinguishers and evacuate immediately. Anytime portable fire extinguishers are used by trained people in an occupancy, the fire department should be called first and the building's fire alarm system (if there is one) should be activated, no matter how small the fire may seem.

NFPA 12 is the *Standard on Carbon Dioxide (CO_2) Extinguishing Systems®*. Carbon dioxide is a colorless and odorless, generally inert gas, which extinguishes fire by reducing the concentrations of oxygen, the vapor phase of the fuel, or both in the air to the extent that the combustion process stops. The same process of oxygen reduction, in which carbon dioxide extinguishes fire, also makes the area dangerous to human occupancy. When discharged into a small room or confined space, carbon dioxide can displace the oxygen in the air and cause simple asphyxiation to persons in the area. In other words, there may not be enough oxygen in the space to support life when the carbon dioxide is released. Therefore, a predischarge alarm is installed to warn occupants of the impending discharge of carbon dioxide. By code, specific warning signs are required for occupants to warn them of the dangers of carbon dioxide when discharged. Carbon dioxide is generally used when an inert electrically nonconductive extinguishing agent is essential or desirable, where cleanup of other media presents a problem, or where it is more economical to install than other systems. CO_2 systems are designed for flammable liquid materials, electrical hazards, engine fires, ordinary combustibles, and hazardous solids. While carbon dioxide will extinguish flammable liquids and ordinary combustibles, it does not prevent reignition if the area around the flammable materials is still hot enough to support combustion. Reignition may occur as a result. Firefighters should wear self-contained breathing apparatus (SCBA) when responding to areas where carbon dioxide has been discharged or ventilate the area prior to entry. Locations of automatic fire extinguishing systems within a facility should be noted by fire prevention personnel during shop drawing review and included on dispatch information for those occupancies. Responding firefighters will then know what type of extinguishing system they may expect to encounter during a response to a particular facility. They will know ahead of time to expect to hook up to a fire department connection to supply a sprinkler system or wear SCBA because the facility has a carbon dioxide system that may have discharged. The more information firefighters have about a facility they are responding to, the quicker they can plan and set up operations when they arrive.

NFPA 12A is the *Standard for Halon 1301 Extinguishing Systems.* Halon 1301 is generally a nontoxic gaseous fire extinguishing agent and a member of the alkyl halide hydrocarbon derivative family with a chemical formula of CF_3Br. The chemical name for Halon 1301 is bromotrifluoromethane and it is also known by the trade names BTM and Freon 13B1. Alkyl halides are made up of carbon and one or more of the halogens, fluorine, chlorine, bromine, and/or iodine. They are also among extinguishing agents that are being phased out by the Montreal Treaty because they have a detrimental effect on the ozone layer and contribute to global warming.

NFPA 13, 13D, and 13R standards provide code information concerning the installation of sprinkler systems, both wet and dry types. NFPA 13 applies to commercial sprinkler systems required by other sections of NFPA codes. It presents the "minimum requirements for the design and installation of automatic fire sprinkler systems and exposure protection sprinkler systems covered within the standard." This standard applies to "character and adequacy of water supplies, selection of sprinklers, fittings, piping, valves, and all materials and accessories, including the installation of private fire service mains." Plans reviewers will use code documents to ensure that sprinkler systems are properly designed, installed, tested, and maintained. Sprinkler systems, while not required in all occupancies at the present time, oftentimes allow for other code requirements to be waived (especially in model building codes). Reviewers should encourage and be prepared to sell the idea of full sprinkler systems in all buildings, including single-family homes.

NFPA 13D applies to "the design and installation of automatic sprinkler systems for protection against the fire hazards in one- and two-family dwellings and manufactured homes." The purpose of the standard is to "provide a sprinkler system that aids in the detection and control of residential fires and thus provides improved protection against injury, loss of life, and property damage." Sprinkler systems designed under NFPA 13D are expected to prevent flashover in the room where the fire starts and to increase occupant escape or rescue potential. NFPA 13R is intended for residential occupancies other than one- and two-family homes. This would include apartments, dormitories, and hotels that are four stories or less in height. Other than the occupancy type, the general purpose of this code is the same as for one- and two-family homes.

Special extinguishing agent systems are covered by other NFPA codes. These systems are used where hazards exist that a wet or dry system will not address or where a wet system would cause damage to expensive or sensitive equipment or priceless items, such as rare books, artwork, and museum pieces. Care should be taken to make sure that a legitimate reason exists for the alternate protection system and that it is being properly used and maintained. Alternate systems should be able to provide the same or greater level of extinguishment protection as a wet or dry sprinkler system.

NFPA 13E, *Recommended Practice for Fire Department Operations in Properties Protected by Sprinkler and Standpipe Systems*, is the standard that provides information on what firefighters should know when responding to facilities that have interior sprinkler systems, exterior sprinkler systems, and standpipe systems. Training programs should be developed using this standard to prepare firefighters to deal with these systems in buildings. Standard operating procedures should be developed as well, and this standard can be used as a resource for those procedures.

NFPA 14 is the *Standard for the Installation of Standpipe, Private Hydrant, and Hose Systems®*. This standard contains requirements for the "installation of standpipe, private hydrant, hose systems, monitor nozzles, hose houses, including methods and procedures of water flow testing for the evaluation of water supplies." Standpipes are a critical element of interior firefighting operations in larger buildings and firefighters should have a good working knowledge of their potential locations and methods of operation. Building owners should be encouraged to develop floor plans that show locations of standpipes and other building fire alarm and fire protection equipment and provide it in a location specified by the fire department to assist them during an emergency (usually near the annunciator panel, if provided). Fire prevention bureau personnel should monitor the installation of new standpipe systems and retrofits to make sure that standpipe valves are installed at the proper locations and at the right angle for easy connection of hose lines. When suppression companies respond to a building for an emergency or when conducting company-level inspections, they too can check to see whether the FDC and standpipe connections are ready for use.

NFPA 15 is the *Standard for Water Spray Fixed Systems for Fire Protection®*. These systems utilize wet spray (mist) and are used for flammable liquid hazards where a wet sprinkler system would not be effective. They are much less messy when discharged compared to dry chemical. Water spray systems provide fire protection against gaseous and liquid flammable materials, electrical hazards, ordinary combustibles, and certain hazardous solids such as propellants and pyrotechnics.

NFPA 17 is the *Standard for Dry Chemical Extinguishing Systems®*. Dry chemical systems are generally used for flammable liquid exposures such as dip tanks and cooking operations using deep-fat fryers or grills. When a fire occurs, dry chemical is expelled from its container by either carbon dioxide or nitrogen, which are inert gases and used only as a propellant; they are not intended to take part in fire extinguishment process. Dry chemical fire extinguishing materials may include, but are not limited to, multipurpose dry chemical (mono-ammonium phosphate) (yellow in color), Purple K® (purple in color) (potassium bicarbonate), and sodium bicarbonate (white in color).

NFPA 17A is the *Standard for Wet Chemical Extinguishing Systems*®. Wet chemical systems are designed as an alternative agent used to protect commercial cooking operations. It is especially useful where an agent is desired that does not require significant cleanup after a discharge of the agent. NFPA 96 further covers requirements for the installation of commercial cooking equipment and related fire protection systems. These systems are required in restaurants where there is deep-fat frying or cooking on a grill or charcoal grill. Response personnel, when making responses to restaurants or fast-food outlets, should check for hood extinguishing systems and current testing and maintenance records. Hood systems are required to be tested and maintained and inspection tags updated to the current month and year of the testing and maintenance.

NFPA 20 is the *Standard for the Installation of Stationary Pumps for Fire Protection*. It covers the "selection and installation of pumps supplying water for private fire protection. Items considered include water supplies; suction, discharge, and auxiliary equipment; power supplies; electric drive and control; internal combustion engine drive and control; steam turbine drive and control; and acceptance tests and operation." Fire pumps are provided in buildings where the local water pressure is not sufficient to move water to the sprinkler system(s) and standpipes within a structure. Buildings equipped with fire pumps may not require the use of fire department pumpers to complete fire extinguishment operations in all cases. However, it is important that in all situations the fire department response provides supply lines and is ready to pump water to the fire department connection if needed in accordance with local standard operating procedures (SOPs). If interior firefighting operations are needed to extinguish a fire, then the standpipe system must be supplied by the fire department to insure proper water volume and pressure to support the firefighting efforts. Most high-rise buildings will have fire pumps, and the height of taller buildings may require the installation of multiple fire pumps to get the water to the upper floors of the building. Fire pumps are required to be tested periodically as required by NFPA 25.

NFPA 22 is the *Standard for Water Tanks for Private Fire Protection*. This standard provides information concerning installation of water tanks for private fire protection. It provides "minimum requirements for the design, construction, installation, and maintenance of tanks and accessory equipment that supply water for private fire protection." Included are gravity tanks, suction tanks, pressure tanks, and embankment-supported coated-fabric suction tanks, towers, foundations, pipe connections and fittings, valve enclosures, tank filling, and protection against freezing. Water tanks are used in locations where the volume of water or pressure in public mains present is not sufficient to supply fire protection systems within a structure or group of structures.

NFPA 24 is the *Standard for Installation of Private Fire Service Mains and Their Appurtenances*. Included in the standard are "minimum requirements for the installation of private fire service mains and their appurtenances supplying automatic sprinkler systems, open sprinkler systems, water spray fixed systems, foam systems, private hydrants, monitor nozzles or standpipe systems with reference to water supplies, private hydrants and hose houses." Private mains are an option for multiple-building complexes where water tanks are not an option and there is no natural static or public water supply to the site available.

NFPA 25 is the *Standard for the Inspection, Testing, and Maintenance of Water-Based Fire Protection Systems.* This standard establishes "minimum requirements necessary for the periodic inspection, testing, and maintenance of water-based fire protection systems. The types of systems addressed by this standard include but are not limited to sprinkler, standpipe, and hose, fixed water spray, and foam water. Included are the water supplies that are part of these systems such as private fire service mains and appurtenances, fire pumps, and water storage tanks as well as valves controlling system flow." Maintenance and inspection of water-based fire protection systems is only preceded in importance by the installation of the systems themselves. Once a system is in place, it should be properly maintained to ensure quick and efficient operation during an emergency. All records from tests conducted need to be maintained on-site for inspection by the authority having jurisdiction.

NFPA 70 and 72 are the *National Electrical Code* and *National Fire Alarm Code*, respectively. The electric code is mentioned here because of requirements for fire alarms and installation, which are located in the *National Electrical Code.* All electric installations in association with fire alarm, fire protection systems, emergency lighting systems, and all other life safety systems should be installed in accordance with the NFPA 70 and any local codes. The *International Building Code* also has an associated electrical code titled ICC Electrical Code. Licensed electricians under the guidance of a fire alarm contractor install fire alarm systems. Once installed, the fire alarm contractor will program and test the system. Many states require that fire alarm technicians be licensed. NFPA 72 covers "the application, installation, location, performance, and maintenance of fire alarm systems and their components." The purpose of NFPA 72 is to "define the means of signal initiation, transmission, notification, and annunciation; the levels of performance; and the reliability of the various types of fire alarm systems." Firefighters should be thoroughly familiar with the components of fire alarm systems and their function as well as methods for determining which detection or activation device has been activated. Because elevators make use of smoke and heat detectors, there are requirements in the elevator code (ASME 17.1) for their installation. Unfortunately, there are also some conflicts between NFPA 72 and the elevator code. These will be discussed further in the fire alarm system chapter.

NFPA 90A and 90B contain requirements for fire and smoke dampers in relation to air-conditioning and ventilating systems as well as warm air heating and air-conditioning systems. NFPA 90A is the *Standard for the Installation of Air-Conditioning and Ventilating Systems.* NFPA 90B is the *Standard for the Installation of Warm Air Heating and Air-Conditioning Systems.* Smoke and fire dampers help prevent and control the spread of smoke, hot gases, and flame through a building via the HVAC duct system. NFPA 90A deals with "all systems for the movement of environmental air in structures that serve spaces over 25,000 ft^3, buildings of

Types III, IV, and V construction over three stories in height, regardless of volume, buildings and spaces not covered by other applicable NFPA standards, and occupants or processes not covered by other applicable NFPA standards." This standard prescribes minimum requirements for safety to life and property from fire. These requirements are intended to accomplish the following:

- Restrict the spread of smoke through air duct systems within a building or into a building from the outside.
- Restrict the spread of fire through air duct systems from the area of fire origin whether located within the building or outside.
- Maintain the fire-resistive integrity of building components and elements such as floors, partitions, roofs, walls, and floor or floor–ceiling assemblies affected by the installation of air duct systems.
- Minimize the ignition sources and combustibility of the elements of the air duct systems.
- Permit the air duct systems in a building to be used for the additional purpose of emergency smoke control.

NFPA 90B "applies to all systems for the movement of environmental air in structures that serve one- or two-family dwellings and spaces not exceeding 25,000 ft^3 in volume or in any occupancy." Smoke and fire dampers should be checked for proper installation and proper location and tested for function as part of final building acceptances for occupancy. During fire damper acceptance tests at the University of Maryland's new high-rise dental school, it was determined that a flaw in the design of dampers was responsible for the failure of many of the damper installations during the functional test. This led to a nationwide recall of that type of damper. Do you test all fire and smoke dampers that are installed? I wonder how many dampers are out there that were not function tested and would not work during a fire.

Elevators are regulated through the American National Standards Institute (ASME) A17-1, *Safety Code for Elevators and Escalators*. This safety code applies to elevators, escalators, dumbwaiters, moving walks, material lifts and dumbwaiters with automatic transfer devices, wheelchair lifts, and stairway chairlifts. Many states and local jurisdictions have elevator inspection departments that inspect elevator installation and maintenance independent of any fire service inspection of elevators. For example, in Maryland, the Department of Licensing and Regulation, Division of Labor and Industry is responsible for elevator inspections. Firefighters are regularly called upon to rescue trapped occupants from inoperative elevators in buildings. They should become familiar with elevator types and operations within buildings in their jurisdiction. Additionally, elevators are equipped with fire service controls so that the firefighters may use the elevators during an emergency. Every firefighter should be familiar with elevator fire service controls and how they work. Access to fire service functions of elevators is accomplished with a key. Firefighters should also make sure keys are available for all building elevators in the event they need to use them during an emergency. Without the key, the elevators will not be operational for firefighting or rescue operations.

FIRE PREVENTION CODES OVERVIEW

NFPA 1 was developed as a result of the requests of members for a document covering all aspects of fire protection and prevention. This document uses other developed NFPA codes and standards for quick field reference in one easy to understand publication. It only contains commonly encountered inspection issues; it does not contain the full text of the cited codes. Some of the topics included are fire drills, smoking, fire lanes, means of egress, building services, fire protection systems, occupancy fire safety, flammable liquids, and others. Fire inspectors and fire companies can take this code with them when conducting field inspections to make quick reference to common issues. If more detail is needed or other issues arise, they can consult the full code when they return to their offices.

The *International Fire Code* is developed and published by the International Conference of Building Officials who also publishes the *International Building Code* and other codes. It is intended to provide information of specific fire issues not covered by the building code. The first edition of the *International Fire Code* was published in 2000 following the work of a development committee appointed by the International Code Council (ICC) made up of membership of the Building Officials and Code Administrators International, Inc. (BOCA), International Conference of Building Officials (ICBO), and Southern Building Code Congress International (SBCCI). Their intent was to draft a comprehensive set of fire safety regulations consistent with and inclusive of the scope of the existing model codes. A new edition is promulgated every three years.

BUILDING CODES OVERVIEW

During 1999, BOCA, ICBO, and SBCCI combined their efforts to form the *2000 International Building Code*. NFPA was also involved in negotiations for one standard building code but could not reach a compromise with the other code-making groups; therefore, NFPA developed their own building code, NFPA 5000, *Building Construction and Safety Code*. In addition to construction requirements, building codes or related fire prevention codes cover fire alarm and sprinkle system requirements. According to the *International Building Code*, its provisions "apply to the construction, alteration, movement, enlargement, replacement, repair, equipment, use and occupancy, location, maintenance, removal and demolition of every building or structure or any appurtenances connected or attached to such buildings or structures." It does not apply to detached one- and two-family dwellings and multiple single-family dwellings (townhouses) not more than three stories high with separate means of egress and their accessory structures. The *International Residential Code* applies to these occupancies. NFPA 5000 addresses "those construction, protection, and occupancy features necessary to minimize danger to life and property." The purpose is to "provide minimum design regulations to safeguard life, health, property, and public welfare and to minimize injuries by regulating and controlling the permitting, design, construction, quality of materials, use and occupancy, location, and maintenance of all buildings and structures within the jurisdiction and certain equipment specifically regulated herein." While the wording is different between the two building codes mentioned, the overall purpose of each code is very similar.

RECOGNIZED TESTING LABORATORIES

Materials used in construction of buildings, along with other fire alarm, fire protection, and fire safety products, need to be tested and labeled by a recognized testing laboratory before they are used in building construction. There are many such organizations nationwide, so only the two most well known will be discussed here. It is important that fire inspectors and plans reviewers are aware of the testing laboratory labels and where to find them on various components of fire safety and building construction components. Fire protection features such as rated walls should be constructed according to a specific system outlined in publications of the testing laboratories. Fire-stopping systems must be installed according to the requirements shown in the testing laboratory publications. Fire walls and fire-stopping systems have detailed descriptions and drawings of how the systems are to be constructed and installed. The detail should appear on construction drawings and a reference made to the listing agency system number, which should be verified by the plans reviewer.

UNDERWRITERS LABORATORIES OVERVIEW

Underwriters Laboratories, Inc. (UL), is an independent, nonprofit product safety testing and certification organization. They have tested products for public safety for more than a hundred years. Over 17 billion products display the UL Listing Marks. Not all of the products listed by UL are fire or building construction related. In terms of fire prevention and inspection, UL lists materials and systems for use in building construction including building materials, fire protection equipment, fire walls, and fire-stopping and related products. The *Building Materials Directory* includes materials evaluated for surface burning characteristics, elevator equipment, exit appliances, emergency power equipment, fireplaces, chimneys, and vents. The *Fire Protection Equipment Guide* contains information on portable fire extinguishers, extinguishing agents, automatic sprinkler and standpipe components, fire pumps and controllers, fire alarm system components, smoke alarms, and fire alarm service providers. The *Fire Resistance Directory*, volume 1, covers hourly ratings for beams, columns, floors, roofs, walls, and patricians. Volume 2 includes fire-rated patricians, through-penetration fire stops, perimeter fire containment systems, and fire-resistive duct assemblies. Volume 3 contains dampers, fire doors, hardware and frames, glass blocks, glazing materials, and leakage-rated door assemblies. UL will provide copies of the above-mentioned directories to AHJs free of charge. Just go to the UL Web site for information (www.ul.com). It is up to the AHJ to determine whether systems or products proposed for individual projects meet the requirements of UL or other recognized testing laboratory.

UL'S MARKS—WHAT THEY LOOK LIKE AND WHAT THEY MEAN

All UL marks are owned by Underwriters Laboratories, Inc., and may not be used by anyone else to label products. There are several types of UL marks, described below. Each has its own specific meaning and significance. The only way to determine if a product has been certified by UL is to look for the UL mark on the product itself. In a few instances, the UL mark may be present only on the packaging of a product.

The UL mark on a product means that UL has tested and evaluated representative samples of that product and determined that they meet UL's requirements. In addition, products are periodically checked by UL at the manufacturing facility to make sure they continue to meet UL requirements. The UL marks are registered certification marks of Underwriters Laboratories, Inc. The UL marks may be only used on or in connection with products certified by UL and under the terms of written agreement with UL.

UL Listing Mark

This is one of the most common UL marks. If a product carries this mark, it means UL found that samples of this product met UL's safety requirements. These requirements are primarily based on UL's own published *Standards for Safety*. This type of mark is seen commonly on appliances and computer equipment, furnaces and heaters, fuses, electrical panel boards, smoke and carbon monoxide detectors, fire extinguishers and sprinkler systems, personal flotation devices like life jackets and life preservers, bullet-resistant glass, and thousands of other products.

The World of Difference between UL "Listed" and UL "Recognized"

A product is UL Listed if the UL Listing Mark is on the product, accompanied by the manufacturer's name, trade name, trademark, or other authorized identification. A UL Listing Mark on a product is always composed of four elements: the "UL" in a circle mark, the word "listed" in capital letters, an alphanumeric control number, and the product name, (e.g., toaster or portable lamp). Sometimes the UL file number is used as company identification. The UL Listing Mark on a product is the manufacturer's representation that samples of that complete product have been tested by UL to nationally recognized safety standards and found to be free from reasonably foreseeable risk of fire, electric shock, and related hazards and that the product was manufactured under UL's follow-up services program. Let us assume, for example, that you are looking at the installation of a spa in a health club. If you can locate a nameplate marking on the spa with the complete UL Listing Mark and the other information noted above, the spa, the end-product, meets the requirements outlined in UL 1563, *Electric Spas, Equipment Assemblies and Associated Equipment*. If you do not find a UL Listing Mark on the product, you may find, on closer examination, that some of the individual components in the spa such as the pump, control, heater, or filter have the UL Recognized Component Mark ℞. And some manufacturers may claim that because the components are UL Recognized, the product in which they are assembled meets all the necessary requirements. But that is not necessarily the case, because the UL Recognized Component Mark means that the component alone meets the requirements for a limited, specified use. Remember, the complete UL Listing Mark and related information on the product indicates that the spa (or other end-product) is UL Listed. UL's Component Recognition Service covers the testing and evaluation of component products that are incomplete or restricted in performance capabilities. These components will later be used in complete end-use products or systems listed by UL. UL's Component Recognition Service covers millions of components, such as plastics, wire, and printed wiring boards, that may be used in either very specific or a broad spectrum of end-products or even components such as motors or power supplies. These components are not intended for separate installation in the field; they are intended for use as components of complete equipment submitted for investigation to UL. Component/end-product compatibility is the critical link between certification of a component and certification of the end-product in which the component is used. Use of UL Recognized Components in a spa (or any other product) does not mean the spa itself is UL Listed. If you are unsure of the exact meaning of a given UL certification (Listing or Recognition), look in the appropriate UL product directory for information about a specific product certification and marking information. For example, the swimming pool and spa equipment category (WABX) begins on page 505 of the *1997 Electrical Construction Equipment Directory*. The directory will also explain any limitations and the extent of UL's evaluation in the information section preceding each product category. If you have exhausted your information sources, here are some ways UL can help. If you have the product name and catalog number, part number or system designation, call UL's data services at 1-847-272-8800, ext. 42396. ULDS will help find the UL category for the product in question. If you need to verify a listing or find a file number for a product bearing a UL mark, call 1-847-272-4909, or customer service at 1-877-ULHELPS. As always, codes and technical services staff members at each UL office will help with other questions you may have concerning UL certifications, code compatibility, or product installation.

C-UL Listing Mark

This mark is applied to products for the Canadian market. The products with this type of mark have been evaluated to Canadian safety requirements, which may be somewhat different from U.S. safety requirements. You will see this type of mark on appliances and computer equipment, vending machines, household burglar alarm systems, lighting fixtures, and many other types of products.

C-UL US Listing Mark

UL introduced this new listing mark in early 1998. It indicates compliance with both Canadian and U.S. requirements. The Canada/U.S. UL Mark is optional. UL encourages those manufacturers with products certified for both countries to use this new, combined mark, but they may continue using separate UL marks for the United States and Canada.

Classification Mark

This mark appears on products that UL has also evaluated. Products carrying this mark have been evaluated for specific properties, a limited range of hazards, or suitability for use under limited or special conditions. Typically, products classified by UL fall into the general categories of building materials and industrial equipment. Examples of types of equipment classified by UL include immersion suits, fire doors, protective gear for firefighters, and industrial trucks.

C-UL Classification Mark

This classification marking is used for products intended for the Canadian marketplace. It indicates that UL has used Canadian standards to evaluate the product for specific hazards or properties. Examples of C-UL Classified products include air filter units, fire-stop devices, certain types of roofing systems, and others.

C-UL US Classification Mark

UL introduced this new classification mark in early 1998. It indicates compliance with both Canadian and U.S. requirements. The Canada/U.S. UL Mark is optional. UL encourages those manufacturers with products certified for both countries to use this new, combined mark, but they may continue using separate UL Marks for the United States and Canada.

Recognized Component Mark ![RU] and Canadian
Recognized Component Mark c![RU]

These are marks consumers rarely see because they are specifically used on component parts that are part of a larger product or system. These components may have restrictions on their performance or may be incomplete in construction. The component recognition marking is found on a wide range of products, including some switches, power supplies, printed wiring boards, some kinds of industrial control equipment and thousands of other products. Products intended for Canada carry the Recognized Component mark "C."

Recognized Component Mark for Canada and the United States c![RU]us

This new UL Recognized Component Mark, which became effective April 1, 1998, may be used on components certified by UL to both Canadian and U.S. requirements. Although UL had not originally planned to introduce a combined Recognized Component Mark, the popularity of the Canada/U.S. Listing and Classification marks among clients with UL certifications for both Canada and the United States has led to the new mark.

Field Evaluated Product Mark

A Field Evaluated Product Mark is applied to a product that is thoroughly evaluated in the field instead of UL's laboratories or the manufacturer's facility. If a product has been significantly modified since its manufacture or the product does not bear any third-party certification mark, a building owner, a regulatory authority, or anyone else directly involved with the product can request that UL conduct tests in the field on the specific piece of equipment. Products that meet appropriate safety requirements are labeled with a tamper-resistant Field Evaluated Product Mark.

Gas-Fired Mark LISTED

UL now offers a new Gas-Fired Listing Mark to be used exclusively on gas-fired appliances and equipment. The Gas-Fired Mark indicates a product's compliance to nationally recognized gas standards, including UL, ASME Z21/Z83 Series, and CSA/CGA standards. The Gas-Fired Mark signals that a product has been evaluated to reasonably foreseeable hazards including both gas and electrical hazards.

Gas-Fired Marks for Canada LISTED

Gas-fired equipment evaluated to Canadian national standards will be authorized to display the Canadian Gas-Fired Mark. For gas-fired equipment evaluated to both U.S. and Canadian standards, the combination U.S. and Canadian mark will be authorized.

UL Verified Performance Mark VERIFIED

The UL Performance Verification Mark can be applied to listed communications cable verified to performance standards such as the UL Performance Category Program, ISO/IEC 11801, and NEMA WC66. It can also be applied to listed fiber-optic cable verified to Telcordia specifications. The UL Performance Verification Mark may also be applied to data transmission or optical fiber cable verified to performance specifications only. The Verified Performance Mark signifies that telecommunications cabling products are certified for both safety and performance and comply with industry performance standards and draft standards. For UL Listing Marks and explanatory information, see http://www.ul.com/mark/

FACTORY MUTUAL OVERVIEW

Factory Mutual (FM) is an insurance company that takes an aggressive approach to risk management. They test materials and systems for their customers from a loss prevention perspective. However, the results of the testing are acceptable to most AHJs for application to other locations. Factory Mutual customers and fire safety professionals rely on FM scientific research and testing capabilities to help them better understand their property hazards and, ultimately, discover better ways to reduce their exposure to these risks. A specialized group of scientists, engineers, and technicians, through ground-breaking research, often sets new standards that advance loss prevention practices and help develop new property loss prevention products. In addition to extensive property loss prevention research, FM offers product certification through an independent, third-party testing laboratory, FM Approvals. Manufacturers rely on FM Approvals to test and approve their products and services, certifying their reliability. Products or services that sufficiently meet FM Approvals' rigorous testing standards may bear the FM Approved mark, a visual symbol of excellence that is widely recognized and respected.

FM Global History

For nearly two centuries, clients have relied on FM Global for ways to prevent, control, and insure against commercial and industrial property losses worldwide.

The Beginning

During the depression of 1835, Zachariah Allen, a prominent textile mill owner, set out to reduce the insurance premium on his Rhode Island mill by making property improvements that would minimize the chance of fire loss. At that time, insurance premium increases for losses were shared among all insureds, regardless of individual loss history.

Although widely accepted today, the concept of loss prevention and control was virtually unheard of at the time. To Allen, a proactive approach to preventing losses made good economic sense. After making considerable improvements to his mill, Allen requested a reduction in his premium but was denied. He called upon other local textile mill owners who shared his loss prevention philosophy to create a mutual insurance company that would insure only "good risk" factories. (Today, these are known as highly protected risks, or HPR properties.) This would result in fewer losses, he reasoned, and hence smaller premium payments. Whatever premium remained at the end of the year would be returned to policyholders in the form of dividends. Sold on the concept, the group agreed and, by year's end, formed the Manufacturers Mutual Fire Insurance Company, the oldest predecessor of FM Global. During the company's first fourteen years, the mill owners and mutual policyholders of Manufacturers Mutual enjoyed an average 50-percent reduction in premium compared with what other insurance companies were charging. As Allen predicted, proper fire prevention methods, monitored by regular fire inspections for mill policyholders, resulted in fewer losses. Despite its initial success, one problem remained for the pioneer mutual insurance company: A single mutual insurance company could not withstand the financial cost of the loss of an entire plant. More capacity was needed, so in 1848, Allen formed another mutual insurance company, Rhode Island Mutual.

Expansion

In 1850, Boston Manufacturers Mutual Fire Insurance Company, the third-oldest FM Global predecessor, was created when Allen convinced a Boston merchant with significant cotton mill ownership to form his own mutual insurance company with like-minded Boston mill owners. Throughout the next twenty years, other mutual insurance companies were added to the group roster. Together, these companies and the ones that later evolved soon became known as the Associated Factory Mutual

Fire Insurance Companies, or the Factory Mutual's, for short. The Factory Mutual's loss record was enviable; losses were less frequent and less severe than those experienced by most other non-mutual insurers. Losses were examined to determine how they were caused and what could have been done to prevent them from occurring. Inspection teams even examined non-policyholder losses to help increase the Factory Mutual's knowledge base. This vital loss information helped identify specific industry hazards and was essential to developing loss control recommendations for policyholders with similar occupancies. Such information was shared among all the Factory Mutual insurance companies and was particularly critical to the inspection teams, which were staffed separately by each individual company. As the FM companies grew, however, the inspection workload became nearly impossible to manage. By 1878, the FM companies formed a dedicated unit to handle the collective inspection activities for all the FM policyholders. This unique group of loss control specialists initially provided just inspection services. The group later began performing appraisals and adjustments, loss analysis, and research activities associated with preventing fire and other hazards—all to benefit the mutual insurance company owners and their policyholders. Today, all of these services remain integral components to FM Global in the form of engineering and research. Like today, the FM companies' main interests in the late 1800s and early 1900s remained focused on researching and developing products or techniques that would help mitigate property risks and advance the efforts of property conservation. In 1874, a revolutionary form of loss control entered the loss prevention scene: the fire sprinkler. While the invention was designed and later perfected by entities outside the FM realm, it was FM's support and promotion of the product that led to its eventual widespread use and acceptance.

The Twentieth Century

The beginning of the twentieth century brought much change for the FM companies. Where once the mutual insurance companies focused primarily on the familiar business of textiles primarily within the northeast region of the United States, new companies began to form that sought business beyond the traditional geographical boundaries. These mutuals began branching out into other industries, such as shoe and rubber manufacturers, foundries and light, and gas and power companies, while still maintaining their preference for HPR properties. During the next seventy-five to eighty years, the need for more comprehensive policyholder coverage grew, forcing a series of consolidations among the FM companies. By 1987, forty-two separate mutual insurance companies had become three: Allendale Mutual Insurance Company, Johnston, Rhode Island; Arkwright Mutual Insurance Company, Waltham, Massachusetts; and Protection Mutual Insurance Company, Park Ridge, Illinois. The three separate organizations found it difficult to deliver competitively priced, value-added engineering services in a marketplace full of increasing competition and a demand for more challenging property protection programs. In 1998, the CEOs announced their intent to merge the three companies to create FM Global. In 1999, the merger was completed, enabling FM Global to leverage the considerable resources and talents of the three former entities to provide consistent, cost-effective, and competitive products for its policyholders around the world.

FM Global Today and Tomorrow

Nearly two centuries after Zachariah Allen's simple premise of making property improvements to reduce risk, FM Global has emerged as an international property insurance and loss prevention engineering leader with US$3.9 billion of in-force premium, US$4.4 billion in policyholders' surplus, and the resources to serve clients in more than one hundred countries. Allen's vision also remains a fundamental principle of FM Global today. We are committed to working in partnership with our clients to reduce threats to their property and operating reliability through state-of-the-art property loss prevention research and engineering and comprehensive insurance products.

2 Arriving on Scene

APPROACHING THE INCIDENT SCENE

Firefighters approaching an incident scene need to be aware of everything going on around the area. Firefighters need to watch for building fire protection devices such as the fire department connection (FDC), fire hydrants, entrances, people who may have evacuated, important signs, Knox Box® locations, and others (figure 2.1). In addition to the usual size-up issues emergency responders need to address at an incident scene, they also need to be alert for criminal activity such as signs of arson, potential acts of terrorism, or hazardous materials when approaching an incident scene. There are important concerns related to criminal activity or potential terrorism when arriving at a building with the fire alarm activated. Watch for obstructions to fire hydrants and fire department connections. Terrorists may be responsible for false alarm activations at target hazard buildings in an effort to observe fire department operational procedures. This surveillance is undertaken as part of the terrorists planning phase in order to determine effective locations to set up explosive or secondary explosive devices or other weapons of mass destruction (WMD) during a terrorist attack. Firefighters should be on the lookout for suspicious persons who may be watching or photographing their operations. Also be alert for suspicious packages or objects when approaching an incident scene. Remember, the firefighter is no longer the good guy in the eyes of the terrorist and may be the target of terrorist activity along with the public, law enforcement, a building, community infrastructure, or a community event. If terrorist activity is to be successful, terrorists must do an effective preplan of the target. This preplan should look at how emergency responders approach a building or target. It should identify where fire personnel park apparatus around the building, staging areas, command post locations, and response routes. Response personnel should not always travel to an emergency at a particular location by the same route, especially those buildings that may be attractive targets to the terrorist. Unfortunately, firefighters are creatures of habit. We almost always take the same route to an incident scene, park in the same location, place our staging areas in the same place, and put our command post in the same location at every incident scene at a particular location. Responding firefighters generally park in front of the building and enter the building through the same doors almost every time they respond to a specific location. That type of routine response can make the planning of the terrorist very easy! Firefighters should take alternate routes to responses at incident scenes, park in different areas at the scene, and have various locations planned for staging areas and command posts. This will help make us less predictable and less vulnerable to a planned terrorist attack. Taking alternate routes to emergencies in buildings can also serve to locate other impediments to emergency response, such as overhead

FIGURE 2.1 This aerial ladder was placed too close to the building upon arrival and soon after it started flowing water operations had to shut down and moved as the fire threatened the position of the ladder.

wires, closed streets, unusual parking or traffic patterns, as well as other potential pitfalls.

Once the fire department arrives at the scene they need a place to park to access the building or fire protection features of the building such as hydrants and fire department connections (figure 2.2). During the planning process of new building construction plans, examiners should identify locations where fire lanes may be needed. Local law enforcement should take steps to ensure that fire lanes and fire hydrants are accessible at all times if needed by the fire department. I cannot tell you how many times I have walked around the campus at the University of Maryland, Baltimore, which is located on the streets west of downtown Baltimore, and found vehicles blocking fire lanes and fire hydrants. Several times these lanes and hydrants have been blocked when the fire department arrived for an alarm sounding. If there had been an actual fire they would have had great difficulty hooking up to the closest hydrants. It never ceases to amaze me that people think the open space by

FIGURE 2.2 Pumper apparatus need to get close enough to the building to facilitate hose line use and to provide access to the fire department connection (FDC).

a fire hydrant is left there for their personal use. The laws are pretty clear, at least in Maryland: no stopping or standing within fifteen feet of a fire hydrant. Yet time after time, UPS, Federal Express, and other delivery vehicles, from soda to compressed gas, are found parking in front of fire hydrants (figure 2.15). Unfortunately, parking in front of fire hydrants is not a high-priority issue with law enforcement. In fact, on several occasions I have found ambulances and police vehicles parked in front of fire hydrants. One day I was walking down the street and found a vehicle from a fire prevention agency parked in front of a fire hydrant. It's no wonder everyone else thinks they can park there as well. Blocked hydrants are not restricted to business districts. In my own neighborhood people constantly park in front of fire hydrants on my block and surrounding blocks.

FIRE HYDRANTS AND FIRE DEPARTMENT CONNECTIONS (FDC)

Two primary external fire protection devices are important for responding personnel to locate when approaching any building. Fire hydrants and fire department connections are vital resources if a fire alarm event turns out to be a real fire. These resources may be overlooked if personnel are not taking fire alarm response seriously. Of course, not all buildings are sprinkled or have standpipes, but unless you already know the building is not sprinkled, you need to look for the FDC (figure 2.2). Fire hydrants are generally under the supervision of the local water department, which tests and maintains the systems in operational condition (figure 2.17). In reality, however, fire hydrants are often left unmaintained or even forgotten in some communities until a problem is encountered when the fire department tries to use them. Hydrant effectiveness in providing water for a fire event depends on an effective maintenance program, accessibility of the hydrant, and water supply to the hydrant. Several years ago I noticed a hydrant on a street corner that had been struck by a car and knocked over. There was no water leak, just a disconnected hydrant. I was afraid someone might take the hydrant and sell it for scrap, so I got a two-wheel cart and took it to my office and contacted the water department to tell them about the missing hydrant and where they could get it back. On several occasions I called them back to see when they would come and get it out of my office and replace the one that was missing. It took months for the water department to replace the missing hydrant, but they never came to pick the one up in my office, so it's still there. It does make a nice conversation piece.

I have witnessed fire hydrants installed with the connections facing away from the street or so close to the ground that you would not be able to attach a suction

Cul-De-Sac   Hammerhead   Alternate Style

FIGURE 2.3 Diagram of Turnarounds.

FIGURE 2.4 Speed bumps or humps can impact fire apparatus access. Due to their suspension, these vehicles must come to a nearly complete stop to pass over these bumps, delaying arrival to a fire scene.

hose to them. Such installations should be reported to the water department to be fixed. It is also not unusual to find hydrants hidden or obstructed by landscaping or fences. In the winter, hydrants can be covered with snow, and residents should be encouraged to keep hydrants close to their homes clear for fire department use. Hydrants can be difficult to see at night, and reflective tape or paint can be effective in making them more visible. Some more progressive communities have placed reflectors (usually blue) in the center of the street indicating the location of nearby hydrants, which is very effective in locating them at night. Firefighters should be aware of the capabilities of the local water system and what general types of fire flows can be expected from hydrants in any given area. The National Fire Protection Association (NFPA) has had a system for quite some time that provides guidance on classifying and marking fire hydrants so responding firefighters will know generally what the fire flow is of the hydrant within a range. Details can be found in NFPA Standard 291 *Fire Flow Testing and Marking of Hydrants*. Table 2.1 shows the hydrant dome colors indicating the rated capacity or fire flow with a minimum residual pressure of 20 psi. Those hydrants with a residual pressure of less than 20 psi should have the pressure stenciled on the dome in black. Hydrant caps should also be painted the same color as the dome. It is recommended by NFPA that the barrel of the hydrant be painted chrome yellow, but other colors can be used depending on local interests. Regular painting of hydrants can also make them more visible. When new fire hydrants are to be installed as part of a construction project, plans review personnel should make sure the hydrants are located where they will not be obstructed by vehicles or other items during normal business operations.

Fire department connections are found in many different locations on the exterior of a building, and some buildings may have more than one connection. High-rise codes require a minimum of two connections unless the local fire department

TABLE 2.1

Classification and Marking of Fire Hydrants

Classification	Rated Capacity (gpm)	Rated Capacity (L/min)	Marking Color
Class AA	>1500	>5680	Light blue
Class A	1000–1499	3785–5675	Green
Class B	500–999	1900–3780	Orange
Class C	<500	1900	Red

approves of using just one. I have also seen buildings that had multiple connections that serviced different systems within the same building. Two of them supplied water to standpipes while another one provided water to a basement sprinkler system. Unfortunately, none of them were marked, so firefighting operations could have been delayed depending on where the fire was located in the building. By code, fire department connections should be marked to indicate the type. There are three general types of fire department connections. Some connections are used for the sprinkler system only, others just for a standpipe system, or they may be combination systems that serve both sprinklers and standpipes. Standpipes can either be wet or dry. If dry, the standpipes depend on the fire department to supply water to fight fires. Wet systems are usually supplied with water from the city main but may be from a static source as well. The building may also be equipped with a fire pump to boost water pressure to get water to the upper floors or most remote portion of the system. Fire department engines can hook up to the standpipes and supplement the system with additional pressure and water volume. Sprinkler systems could be installed to only cover certain floors of a building. That information should also be provided on a sign above the fire department connection.

Fire codes require that the FDC be located within a specific distance from the closest fire hydrant. NFPA 14, *Standard for the Installation of Standpipe and Hose Systems*, under section 4-3.5.4 requires the fire department connection for each standpipe be located no farther than one hundred feet from the nearest fire hydrant connected to an approved water supply. Codes also dictate the mounting height of the FDC on the building so that it is accessible to the fire department. NFPA 14 requires that the FDC be located not less than eighteen inches or more than forty-eight inches above the ground. Most codes require that the FDC be on the street side of a building and fully visible from the street or nearest point of fire department access to the building.

Fire departments should provide input to building designers to make sure that the FDC is installed where they want it to be on a building. However, even with this requirement, connections sometimes still get placed on loading docks, behind landscaping, or in places where they can be blocked by vehicles or trash dumpsters. Landscaping starts out small when it is first installed but can grow over the years and may cover the FDC, making it difficult for firefighters to locate and use (figure 2.18). Plants should not be allowed in front of fire department connections

FIGURE 2.5 Fire lane signage is important, both for the public and enforcement officials.

no matter how small they are. Fire prevention and plans review personnel need to tell contractors and building designers where they want the connections located and follow up to make sure they are properly installed. Buildings under construction are required to have a temporary standpipe system installed for the fire department to use if a fire occurs during construction prior to the building's permanent system being installed. Locations of these temporary standpipes are required to be clearly marked with signs and kept clear of obstructions. Fire inspection and plans review personnel should make sure the contractor is aware of the requirement and conduct field inspections to make sure it has been installed.

Codes require that nothing be placed in such a manner that hose lines cannot be attached to the inlets of the fire department connection or obstruct FDCs. This includes features of the building construction as well as landscaping. When hoses are connected to a fire department connection they should not block exits from the building. Locations of the FDC in respect to means of egress should be evaluated during plans review. Fire department connections are another fire protection item that responding companies should look for when the fire alarm is sounding in case they need to use it on this or a future alarm. Fire prevention bureaus should determine what location is best for the local fire department and maker sure that the FDCs are placed accordingly. During installation of fire department connections, no shutoff valves are permitted to be placed between the FDC and building sprinkler and standpipe systems. Buildings with fire pumps will have fire pump hose test connections located on the outside of a building for periodic testing of the pumps. These connections are only used for testing the fire pump and have no fire department function under normal circumstances. However, a fire pump in one building could certainly be used to fight a fire in another building when the scope or conditions of an emergency warrant it. Having a fire pump is like having an engine company in the building. Hoses could be connected to the fire pump test header (discharge) and the pump turned on to supply water to hose lines to fight a fire in another building or

FIGURE 2.6 Fire lane curb has been painted red indicating that no parking is allowed. In this case the fire lane is to protect building access for the fire department apparatus.

outside the building with the fire pump. This is not considered a common procedure but could certainly be used in an emergency situation, especially if a building has an independent water supply. Fire pump test connections may have valves installed or they may be removed, which makes them resemble an FDC. The major difference between FDCs and fire pump test headers is FDCs have female fittings and are designed for the intake of water, and fire pump test headers have male fittings and are designed for discharge of water. However, when the valves are removed, both will have female connections. Firefighters must then rely on the markings on the connections to determine what their purpose is to avoid delays in getting the water supply for fire suppression activated.

The number of ports or valves on the fire pump test header is an indication of the gallons per minute rating of the fire pump (figure 2.9). Each separate connection or valve accounts for 250 gallons per minute (gpm). Therefore, a fire pump test header with two valves would indicate a 500-gpm fire pump, and four valves would indicate a 1000-gpm fire pump. Seeing the fire pump test header on the exterior of the building is a visual indication to responders that the building is equipped with a fire pump.

The Knox Company, known for years for the Knox Boxes® used to safely store building keys for fire department use, has developed locking fire department connection caps to protect threads from damage and to keep the openings from being stuffed with trash or other obstructions (figure 2.20). Normally, brass caps are used to cover openings to the FDCs. Brass caps are often stolen and sold for scrap by vandals and homeless people. Cheaper aluminum caps are often used to replace the more expensive brass caps. They too are removed for scrap. Plastic caps are used as a last resort because they have little if any scrap value. Vandals may remove the

FIGURE 2.7 Fire lane with painted curb and "No Parking" stenciled on the curb.

plastic covers as well (figure 2.22). Knox® Caps keep people from removing brass and aluminum caps for salvage or the plastic caps that are sometimes used to replace the aluminum (figure 2.21). The Knox Caps are placed inside female threads of the FDC and locked in place with a special key. Fire department companies carry the special keys to access the FDC when needed. Using this type of cap reduces the maintenance requirements for the FDC and protects the FDC from vandalism.

FIGURE 2.8 Key boxes (also called "access boxes" or "lock boxes") are small lockable vaults mounted outside building entrances with keys to the building locked inside.

FIGURE 2.9 In multi-tenant buildings, such as shopping centers and malls, tenants usually have rear exit doors that firefighters may access. Often these doors look alike, making it hard to correlate a given door with a particular tenant. Labeling rear doors to the outside with the tenant's name, address number and/or suite number prevents this problem.

BUILDING ACCESS

Access to a building and its fire alarm and protection system is important for the quick mitigation of incidents that may occur there. The faster the fire service can respond, enter, locate the incident, and safely operate in a building, the sooner they can mitigate an incident in a safe manner for themselves as well as occupants. Upon arrival on scene, one of the first priorities is to properly position fire apparatus, which can be critical to operations at a fire scene. In particular, placing aerial apparatus is important for positioning of the aerial ladder or elevating platform (figure 2.1). Pumper apparatus also need to get close enough to the building to facilitate hose line use and access to the fire department connection (figure 2.14). Many structures are situated on public streets that provide firefighting access. Others, which are set back from public streets, have private fire apparatus access lanes, or fire lanes for short. These enable fire apparatus to approach the building and operate effectively. Fire lanes can be dedicated to fire services use or can serve ordinary vehicular traffic as well.

When designing a new building or changing occupancy of an existing building, there are many considerations for both public roads and fire lanes: clear width, clear height, length, turn radius, arrangement, distance from the building, and paving materials all need to be evaluated. In all cases, the most stringent practicable dimensions should be considered for design, since future apparatus purchases or mutual aid apparatus from other jurisdictions may exceed the specifications required in a given jurisdiction at any given time. Minimum building access for fire apparatus is a function of the access road reaching to within a certain distance of all portions of the building's first floor exterior walls. This limit in NFPA 1 and the IFC is 150 feet for buildings without a complete sprinkler system. For fully sprinkled buildings, NFPA 1 permits this distance to be increased to 450 feet; the IFC leaves this decision up to

FIGURE 2.10 Any door that appears to be functional from the outside, but is unusable for any reason, should have a sign reading "This Door is Blocked."

the discretion of the code official. Further, NFPA 1 requires that the road extend to within fifty feet of an exterior door providing interior access. The distance from the building to a road or fire lane is sometimes referred to as "setback distance." NFPA 1141 has additional guidelines for access locations versus building location, with variations depending upon building size, height, sprinkler protection, and separation from other buildings. The options available for attacking a fire increase as more of a building's perimeter becomes accessible to fire apparatus. A concept known as "frontage increase" appears in the IBC and NFPA 5000. If a structure has more than a certain percentage of its perimeter accessible to fire apparatus, these codes allow the maximum size of the building to be increased. Ideally, the full perimeter would be accessible.

A single access route is a basic requirement in both NFPA 1 and the IFC. However, both codes allow the code official or authority having jurisdiction (AHJ) to require additional access routes due to various factors that could inhibit access (such as terrain, climate, or vehicle congestion). NFPA 1141 requires two access routes for buildings over two stories or thirty feet in height. Multiple fire lanes should be as far removed from one another as practicable. Long, dead-end fire lanes or roads should provide a means for fire apparatus to turn around. Both NFPA 1 and the IFC require turnaround space for dead-ends that are more than 150 feet long. There are a number of configurations that facilitate turning maneuvers. These include, "T-turn," "Y-turn," and round cul-de-sac–style arrangements.

NFPA 1141 requires a 120-foot turnaround (figure 2.3) at the end of dead ends more than three hundred feet long. The basic clear width requirement for apparatus access in the IFC and NFPA 1 is twenty feet. NFPA 1141 calls for one-way fire lanes that are sixteen feet wide; however, this applies to roads that do not abut buildings. A clear width of twenty feet will allow most aerial apparatus to extend the outriggers

FIGURE 2.11 Large, unusual, or complex buildings present a challenge to maneuvering and locating specific areas. Directional signs with room/tenant numbers, and graphic directories of tenant/agency layout can assist the public.

FIGURE 2.12 University Police at the University of Maryland Baltimore are on duty 24/7 and provide access to buildings when the fire department responds. So instead of placing Knox Boxes outside of buildings with keys we place them in the Fire Command Center or next to the Annunciator panel inside the building with keys for the inside of the building.

necessary to support the aerial ladder or elevating platform while in operation. However, some recently manufactured aerial apparatus require twenty-five feet of clear width for outrigger deployment. Height is also an issue for fire lanes allowing fire department access for buildings. NFPA 1, NFPA 1141, and the IFC require 13 feet 6 inches as a minimum. Some modern aerial apparatus may require fourteen feet of clearance. NFPA 1 requires at least fourteen feet in colder climates where snow and ice may accumulate on the road surface reducing the height. Overhead wires and other obstructions should be avoided when planning fire lane locations.

Security concerns may impact fire service access. When approaching the scene, be aware of manual or automatic gates that restrict access to a property. Manual gates cause inherent delays because personnel must dismount to unlock them or cut through chains. However, they can also help keep the fire access lane clear by preventing vehicle parking. Responding firefighters should be provided with keys or access cards to any locked or restricted areas. Knox® Boxes work well to provide a location for fire department access. During the design process, be aware of the impact of response by gates, bollards, pop-up barricades, and other perimeter controls that may delay fire service operations. Carefully coordinate between those responsible for security and fire protection, which may help resolve concerns of both. Additionally, proper gate size, location, and swing can facilitate fire service access. Speed bumps or humps can impact fire apparatus access. Due to their suspension, these vehicles must come to a nearly complete stop to pass over these bumps, delaying arrival to a fire scene (figure 2.4). Dips should also be avoided so that long-wheel-base vehicles do not hit bottom and damage undercarriage components and overhanging equipment.

Fire lane signage is important, both for the public and enforcement officials (figure 2.5). Examples include signs, curb painting (figure 2.6), or curb stenciling (figure 2.7). A jurisdiction's requirements must be followed exactly to ensure that

FIGURE 2.13 Buildings may be posted with signs using the NFPA 704 marking system. This identifies certain types of hazardous materials which may be present in a given building.

FIGURE 2.14 Firefighters need to watch for building fire protection devices such as the FDC, fire hydrants, entrances, people who may have evacuated, important signs, Knox Box® locations and others when approaching the scene.

no-parking provisions are legally enforceable. Speed bumps should be conspicuously painted and signs indicating their locations should be posted in climates subject to accumulation of snow and ice. Load limits should be posted conspicuously on both ends of bridges or elevated surfaces. Fire service personnel must be able to rapidly identify and locate a specific building. Address numbers should be placed on the building facing the street or road on which the building is addressed. If the building entry faces a different street, both the street name and the number should be on the

FIGURE 2.15 Fire hydrant access located across the street from major medical school hospital obstructed by United Parcel Service vehicle and police car.

FIGURE 2.16 Not all buildings are sprinkled or have standpipes but unless you already know the building is not sprinkled you need to look for the FDC when approaching the scene.

address sign. Numbers should be large enough to read from the street or road. If this is not possible due to the location of the building or due to obstructions, additional signs should be provided. The IFC specifies that address numbers be a minimum of four inches high. Some jurisdictions have a minimum height requirement, especially for commercial properties. The number should be in Arabic numerals rather than spelled out (for example, "120" instead of "One Hundred Twenty"). Whenever possible, signs should be illuminated. In areas subject to snow accumulation, signs should be positioned above anticipated accumulations.

Once firefighters have arrived and positioned their apparatus, they must go to work. Some factors affecting their efficiency include: the distance and terrain between the apparatus access and the building; how easily they can enter the building; the building's interior layout and vertical access (stairs/elevators/roof access); and how quickly firefighters can locate fire protection features and utilities. The designer can make a positive impact in all of these areas. Both the IFC and NFPA 1 contain requirements for access to building openings, such as approved walkways that lead from the apparatus access points to the entry doors. Fire department pumpers carry hose lines for attacking fires. These are usually smaller in diameter than the hose lines used to supply water to the pumper from a water source. Many pumpers have one or more hose loads of a fixed length connected into a pump discharge. These are "preconnected" hose lines, often called simply "preconnects." Firefighters deploy them rapidly for quick fire attack. However, their useful range is limited by their length, which is generally between one hundred and four hundred feet. Designers planning unusual designs for their buildings or working with unusual sites should coordinate with the local fire department regarding hose line access unless a standpipe system is provided in the building.

Buildings under construction or renovation pose their own particular concerns to the fire service. Code provisions for buildings under construction can be found in IFC chapter 14, IBC chapter 33, NFPA 5000 chapter 14, and NFPA 241. Designers should consider the accessibility of the fire department connections, fire hydrants, and entry points. Consideration also needs to be given to site access for apparatus. The International Fire Code requires "approved" vehicle access and the road must extend to within one hundred feet of temporary or permanent fire department connections. NFPA 1, the *Uniform Fire Code*, requires access roads be provided and maintained during construction. Access roads must reach within fifty feet of a single exterior door that provides access to the building's interior. Some locations may be more likely to be obstructed by construction storage, truck unloading, cranes, phasing of the construction, and security fences. Designers should consider specifying these locations and the location of temporary and permanent fire protection equipment to avoid conflicts. If firefighters need to conduct interior fire suppression operations, they must enter the building at one or more points. The fire service has an array of tools to force entry into buildings. However, forcing entry takes extra time and usually damages the building. Key boxes (also called "access boxes" or "lock boxes") are small lockable vaults mounted outside building entrances (figure 2.8). They are opened with a master key held by the fire department. Inside the box are the building's keys. Some jurisdictions require key boxes; others give building owners the option of installing them or risking the need for firefighters to force entry into their buildings, along with any resulting damage. Code officials enforcing the IFC and NFPA 1 may require key boxes. When key boxes are optional, designers may want to educate owners on their benefits.

First arriving firefighters will often base their point of entry on which windows have fire or smoke venting from them. In most cases, entrances that service any particular window will be readily apparent from the outside. If it is not obvious which door to enter to reach which area, signs or diagrams should be provided

FIGURE 2.17 Fire hydrants are generally under the supervision of the local water department who test and maintain the systems in operational condition.

FIGURE 2.18 Landscaping starts out small when it is first installed but can grow over the years and may cover the FDC making it difficult for firefighters to locate and use.

outside each entrance door indicating portions of the building accessible from the corresponding door. In multi-tenant buildings, such as shopping centers and malls, tenants usually have rear exit doors that firefighters may access. Often these doors look alike, making it hard to correlate a given door with a particular tenant. Labeling rear doors to the outside with the tenant's name, address number, and/or suite number using lettering at least six inches high with a ½-inch stroke (thickness of lines in each letter) prevents this problem (figure 2.9). Any door that appears to be functional from the outside but is unusable for any reason should have a sign reading "This Door is Blocked" (figure 2.10). The lettering should be at least six inches high with a ½-inch stroke. If these doors are properly marked, firefighters will not waste time trying to gain entry through them.

Large, unusual, or complex buildings present a challenge to maneuvering and locating specific areas. Directional signs with room/tenant numbers, and graphic directories of tenant/agency layout can assist the public (figure 2.11). The same diagrams may assist firefighters if they include: stairway and elevator identifiers, fire hose valve locations, fire alarm control panel location, fire alarm annunciator location, fire pump location, and other fire protection features. Diagrams should also contain features to assist unfamiliar users with orientation, such as road names or a compass point. Detailed floor plans showing building layout and fire protection systems can assist the fire service. In buildings with fire command centers, a good location for these plans is in this command center. In other buildings, these plans may be locked inside the fire alarm annunciator panel. A routine function in any advanced fire suppression operation is to control (usually shut down) utilities. Making it easy to locate and identify utilities will speed firefighters' progress. Electric, gas, and other fuel controls should be located either in dedicated rooms with exterior marked entrances or at exterior locations away from openings such as windows or doors. NFPA 170, *Standard for Symbols for Use by the Fire Services*, contains

FIGURE 2.19 The number of ports or valves on the fire pump test header is an indication of the gallons per minute rating of the fire pump.

symbols for marking gas and electric shutoffs. Air-handling equipment should also be prominently marked, especially if located out of sight. The fire service may need to quickly access rooms containing the following equipment: water service, control valves, fire pumps, electric service, switchgear, generators, fans, and other

FIGURE 2.20 The Knox Company, known for years for the Knox Boxes® used to safely store keys for fire department use, has developed locking fire department connections Caps to protect threads from damage and to keep openings from being stuffed with trash and other obstructions.

mechanical equipment. Lettering for this signage should be at least six inches high with a ½-inch stroke (thickness of lines in each letter), unless the standard symbols are used.

Responding to fire alarm activations, because of their frequency, can become a mundane experience for firefighters. A high percentage of reported fire alarm soundings turn out to be false alarms, furthering the risk of complacency. Because of the frequency of false fire alarms, firefighters may not take responses seriously, especially when they occur frequently at certain buildings. Complacency may lead to mistakes if firefighters fail to maintain an operational edge despite the frequency of fire alarm activation calls. Recently an alarm sounded at approximately 11:30 p.m. in a ten-story research building at a major university. Firefighters arrived and multiple flow alarms were illuminated on the fire alarm annunciator panel. University police and maintenance personnel met firefighters at the main entrance to the building and led them to the annunciator. A determination was made that it must have been a false alarm because so many flow switches were illuminated. It is unclear whether building personnel led them to believe it was another false alarm or not. Firefighters chose not to search the building and left maintenance personnel to reset the fire alarm system. However, a sprinkler head had activated on the eighth floor of the building in response to a fire in a research lab. The fire department and building maintenance staff assumed that it was another false alarm because so many flow switches had activated. They left the scene without checking the building and did not find the fire. The sprinkler system ran for several hours, causing a flood with extensive damage on three other floors. It's a good thing that the building was equipped with Early Suppression Fast Response sprinklers that extinguished the fire even without the fire department! I am sure this is not the only time such an event has occurred in the fire service here in the United States. Firefighters should have a thorough understanding of the workings of fire alarm systems and what types of information the systems can and cannot tell them. They also need to be able to read between the lines of what the panel is reporting. Each response to a fire alarm should be treated as a potential fire event. Buildings should be thoroughly checked out before the fire department leaves the scene. The process of checking out the building also provides firefighters with an opportunity to become familiar with the layout of the building and locations of fire protection and fire alarm system controls.

Fire alarm activations that turn out to be false most often involve only one device that has sensed some material other than smoke or has malfunctioned. If multiple devices are activated, it is much more likely that the alarm is not false and certainly warrants further investigation. Another incident that occurred at the same university mentioned above involved activation of the fire alarm system in another building. When the fire department arrived, a flow alarm was illuminated on the annunciator panel for the second floor of a seven-story combination research and classroom building. In addition to the flow alarm activation, the annunciator also indicated that the fire pump was running. This time, personnel searched the second floor and found nothing, so they left. Once again, it turned out that a sprinkler head had activated on the fifth floor from a fire that was set. The flow switch on the fifth floor malfunctioned and the second floor flow switch was the next to activate. Building maintenance personnel noticed the fire pump was running and thought that was an important event and decided to check the building further. They discovered an extinguished fire and considerable water on the fifth floor of the building. These examples of misunderstood incidents illustrate why first responding firefighters should have knowledge of the workings of fire protection systems and other building systems so they can do a better job determining what has occurred when they arrive at a building with a fire alarm activated.

Firefighters can take advantage of fire alarm and other responses to buildings to become familiar with the facility. If a real fire or emergency were to occur, they would be better prepared to effectively deal with the emergency. When arriving at a building, take the time to determine where fire department connections (FDCs), fire alarm panels, sprinkler valves, fire pump rooms, standpipe connections, emergency shutoff valves for utilities, and other life safety features are located within a facility. Make it a part of a preplan or have the information entered into dispatch computer systems so it is printed out when a call comes in. While at a particular facility, take a few minutes to do some spot checks for emergency procedures such as fire evacuation plans, posting of evacuation floor plans, and locations of emergency access keys (or access cards where card readers are used). Each building should have available a set of keys for each elevator and keys to all locked rooms in the building to enable fire personnel to make a thorough building search. Multiple elevator keys are necessary in the event the fire department needs to use more than one elevator at a time. Keep in mind, though, that if the power goes off in the building, usually only one elevator will be on emergency power, which will restrict elevator use by the fire department. Where there is a generator and only one elevator can be operated at a time, there will be a selector switch in the elevator lobby or fire command center that allows firefighters to select which elevator will be on emergency power. It is important that plans reviewers take note of buildings with generators and require the generator to be big enough to handle the operation of all elevators in the building simultaneously if needed. Not all buildings may have emergency power. Keys designated for fire department use can be placed in a Knox® Box outside the main entrance for easy fire department access. Where I work at the University of Maryland, Baltimore, university police are on duty 24/7 and provide access to buildings when the fire department responds. So instead of placing Knox Boxes outside the buildings with keys we place them in the fire command center or next to the annunciator panel inside the building (figure 2.12). This provides the fire department with easy access to the keys when they arrive. Responding to fire alarms in buildings can be turned into an educational experience for fire department personnel. Taking the extra time to review features of the building may lead to more effective operations should a real fire event or other emergency occur.

Copies of certain local fire codes, such as the National Fire Protection Association (NFPA) 1, *Uniform Fire Code*, the *International Fire Code* (IFC), or other fire codes should be made available to responding companies to answer spot questions about life safety systems, building hazards, or common code issues that come up during a response to a building. The fire code is primarily a compilation of most commonly encountered code issues from other areas of the codes. It is written in

FIGURE 2.21 Knox® Caps also keep people from removing the aluminum caps for salvage that are sometimes used to cover FDC openings.

a format that allows field personnel to quickly reference code sections to answer questions that come up during responses or company level inspections. In this author's opinion, NFPA 1 is easier to navigate and carry around in the field. But once you are used to your own fire codes, they can be useful as well. But the key is familiarity.

The University of Maryland, Baltimore, provides fire department information packets next to each annunciator panel at the main entrance to campus buildings or in fire command centers. Building information packets are designed to aid response personnel in navigating the building and finding fire protection and other emergency features. Packets contain bound laminated sheets of building floor plans showing the following information if applicable to a specific building:

- Sprinklers
- Fire pump and its size
- Roof access
- Roof hydrant
- Building construction type
- Building size in stories
- Emergency power
- Elevator phases
- Fire department connection
- Fire department standpipe connections
- Annunciator panel
- Fire alarm panel
- Elevator and other keys
- Fire pump location
- Sprinkler valve locations

FIGURE 2.22 Plastic caps are sometimes used to cover openings in the FDC instead of the aluminum caps because they are not likely to be removed for salvage. However, they are still subject to vandalism.

Information is also provided for locations of special building functions or locations such as animal research facilities, hazardous materials, radiological materials, and biological hazards that may provide challenges for firefighters. Binders are also being developed to include the packets of all university buildings to be given to first due companies and chief officers for access en route to the university during an emergency response. Fire department information will be provided as well on the university's Environmental Health & Safety (EHS) Website on the "Fire–Life Safety Page" (www.ehs.umaryland.edu/firesafety) for quick computer access.

Buildings may also be posted with signs using the NFPA 704 marking system (figure 2.13). This system identifies certain types of hazardous materials that may be present in a given building. NFPA 704 is a voluntary system and is not required unless adopted by the local code authority. The system uses a diamond-shaped sign with four colors, one in each quarter of the sign. Red signifies flammability hazard, blue health hazard, yellow reactivity hazard, and white special information. Numbers are placed inside each quarter of the diamond and range from zero to four, zero indicating no hazard and four indicating the most serious hazard. The signs do not identify individual chemicals but rather hazard classes. It is pretty much a stop sign for first responders, telling them that there are hazardous materials in the building and they need to get more information before taking mitigative actions.

ANNUNCIATOR PANELS

Once the fire department has arrived at a building with a fire alarm sounding and has located the FDC and fire hydrants, the next task is to determine why the fire alarm has been activated. There are two devices within a building that will accomplish this task. One is the annunciator panel and the other is the fire alarm panel. Not all buildings will have both or they may have abbreviated versions of annunciator panels

FIGURE 2.23 Fire alarm panels are usually located in electrical rooms in mechanical areas in basements and penthouses far away from the building entrance.

depending on the age of the building. The annunciator panel (figure 2.23) is usually located just inside the main entrance of a building because of the generally remote location of the fire alarm control panels (figure 2.24), which are the "nerve center" of the fire alarm system. When a building has a fire command center, the annunciator will be located inside along with the fire alarm control panel (figure 2.25). However, if the location of the FCC is not obvious, a second annunciator should be installed at the main entrance. Buildings with multiple entrances the department could respond to may have more than one annunciator. The annunciator is designed

FIGURE 2.24 The annunciator panel is usually located just inside the main entrance of a building.

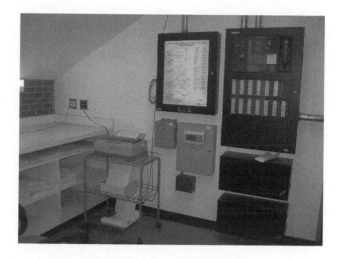

FIGURE 2.25 When a building has a fire command center, the annunciator will be located inside along with the fire alarm control panel.

to give emergency personnel a "mirror image" of what is happening at the fire alarm control panel. There are usually no fire alarm controls at the annunciator. The fire alarm annunciator primarily consists of building floor plans and an arrangement of light-emitting diodes (LEDs) to indicate what alarm device is in alarm and where it is located in the building. Some annunciator panels also contain or have located in another panel next to them a textual message readout (figure 2.26). This readout indicates the location (address) and type of device activated. Older analog systems

FIGURE 2.26 Some annunciator panels also contain or have in a separate panel next to them a textual message readout. This readout indicates the location (address) and type of device that has been activated.

may not be able to distinguish what type of device has gone into alarm and will only display the floor where the alarm has been activated. The annunciator panel is often the first indication to firefighters what the emergency may be within a building. Some buildings will have modern annunciators with detailed floor plans and LEDs. Others will have older ones with no lights and limited floor plans and some systems will not have an annunciator at all. The annunciator panel contains information sent from the building fire alarm control panel. If the fire alarm system devices are addressable, the annunciator will give an indication of the device type along with floor and zone. If the system is not addressable, it will likely only give the floor and/or zone. Older analog fire alarm systems are hard-wired and have little capability for reprogramming duct detectors and elevator machine room devices to supervisory signals. Fire alarm control panels are usually located within mechanical or electrical rooms of buildings. These rooms are required by code to be locked to control access to the fire alarm control panel and power shutoffs. They are often located in the basement or penthouse of a building but may also be located on the individual floors.

Three types of signals can be transmitted by a fire alarm panel and displayed at the fire alarm annunciator; they are supervisory, trouble, and fire alarm. Firefighters are only called when the system is in fire alarm mode. However, make note of any trouble or supervisory alarms that may be active on the panel when you arrive. Supervisory signals can indicate a sprinkler valve is turned off, which could mean that a portion of the building may be without sprinkler coverage. Supervisory signals may also indicate a duct detector, elevator machine room smoke, or heat detector or beam detector. Sprinkler system status would be an important bit of information if the building is on fire. Fire alarm signals are the only ones that summon the fire department. All other fire alarm panel signals are intended for maintenance or alarm company personnel to alert them to a problem with the fire alarm system. Law enforcement may be summoned for supervisory signals where sprinkler valves have been shut off. Supervisory signals are a higher priority than trouble signals and must be responded to by security, maintenance, or repair personnel within two hours of receipt of the alarm. Supervisory signals are generally for a fire alarm system device that requires immediate attention, such as tamper switches on sprinkler valves, duct smoke detectors, elevator machine room and shaft detectors, and others determined by the authority having jurisdiction to not require a general fire alarm signal. Trouble alarm signals must be responded to and addressed by maintenance or repair personnel within four hours after receipt of the alarm. Trouble signals are indications that something is not working right on the fire alarm system, such as broken wires, defective devices, shorts, and others. Trouble and supervisory signals usually sound a local alarm at the fire alarm control panel and may also sound at the annunciator panel. Make sure that building owners know they must respond to trouble and supervisory signals as required by code if they are active when you arrive.

AREAS OF RESCUE ASSISTANCE

Handicapped persons in buildings who rely on the building elevator(s) for access during normal operations will not be able to use the elevator(s) when the fire alarm sounds in the building (figure 2.27). Elevators are dangerous under fire conditions

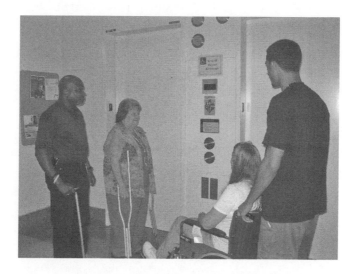

FIGURE 2.27 Handicapped persons who rely on the building elevator for access will not be able to use the elevator when the fire alarm sounds.

or when the fire alarm has been activated because you have no way of knowing whether it is a real fire or not and therefore they should not be used by any building occupants. Elevators may also be recalled by the building fire alarm system, which would prevent occupants from using them anyway. Provisions need to be made to make sure that those who rely on elevators know what to do and are sent to an area of safety until firefighters can remove them from their floor in the building. These safe locations are sometimes referred to as "areas of rescue assistance." Since firefighters will rescue persons who cannot evacuate on their own, it is also important that firefighters be made aware of the locations of areas of rescue assistance in buildings (figure 2.28). Signs can be placed in the fire command centers or next to building annuciators if there is no fire command center. Locations can also be marked on evacuation floor plans throughout the building (figure 2.29). Additionally, signs will be placed at the areas of rescue assistance with the universal handicapped symbol. Buildings that are fully sprinkled do not require specially constructed areas of rescue assistance. Any place on a floor can be designated. If buildings are not sprinkled, then areas of rescue assistance must be fire rated and have communications systems present for handicapped persons to use. Stairwells can be designated as areas of rescue assistance provided that they have been constructed with space available for persons in wheelchairs on landings without obstructing egress of others using the stairs. At the University of Maryland, Baltimore, for many years we used the stairs as areas of rescue assistance. However, after talking with handicapped persons and evaluating our procedures, we have decided to change the locations to main elevator lobbies. All of our buildings are sprinkled, so the code allows us to place the area of rescue assistance anywhere on a floor. The fire department is likely to use the elevators to evacuate persons who need evacuation assistance. Having the people already in the elevator lobby rather than waiting in stairwells will save time

FIGURE 2.28 It is important that firefighters be made aware of the locations of areas of rescue assistance in buildings because they will have to rescue occupants from these locations.

in the evacuation process. Firefighters do not have to check every stairwell on every floor to find those persons needing assistance because they will already be in the elevator lobby. Using the elevator lobby also removes the feeling of isolation that occurs with persons waiting in the stairwell.

Evacuations from buildings need to be taken seriously by building occupants. To help ensure complete evacuation, we use a system of fire wardens at the university.

FIGURE 2.29 Locations of areas of rescue assistance can also be marked on evacuation floor plans throughout the building.

Fire wardens are appointed and trained for each floor of each occupied building. Each warden goes through a ninety-minute training program and is issued an orange fire warden badge. Their job is to assist in the evacuation of their floors, direct persons requiring evacuation assistance to areas of rescue assistance, and make a report to the building evacuation supervisor. Wardens will search all areas of their floors to make sure everyone has evacuated. If there are people who have not evacuated, the fire wardens will invite them to leave. If following the invitation to leave people still refuse, their names and locations will be noted and administrative action will be taken. On some occasions, university police will remove people who refuse to evacuate on their own. The first fire warden who reaches the lobby of the building becomes the building evacuation supervisor by default. They obtain the fire warden clipboard, which has fire warden report forms attached, from the security guard desk. The forms prompt the building evacuation supervisor to obtain information from the reporting wardens that will then be reported to the firefighters when they arrive. If a fire alarm response turns out to be a real fire or emergency, firefighters will evaluate the situation and determine whether persons in areas of rescue assistance need to be evacuated from the building. With the fire warden system in place, firefighters will know floor conditions and locations of persons awaiting rescue when they arrive.

FIRE COMMAND CENTER

High-rise buildings are required by the building and fire codes to have a fire command center (FCC) for the use of response personnel during an emergency (figure 2.30). High-rise buildings require that a voice-type fire alarm system be installed in the building rather than the usual fire alarm horns or bells. Voice messages can be customized to give building occupants special information or directions such as location of exits. There is also an option with high-rise fire alarm systems to do selective evacuation known as "three-floor evacuation." This usually involves the evacuation

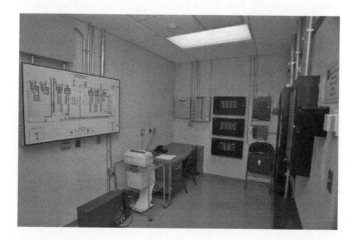

FIGURE 2.30 High-rise buildings are required by the building and fire codes to provide a fire command center (FCC) for the use of response personnel during an emergency.

of the floor where the fire alarm device is activated along with the floor above and the floor below. This reduces the disruption of building operations to other floors during fire alarm activations in the building. Three-floor evacuation works great for false alarms and drills. However, the downside of the three-floor evacuation is that it can cause confusion when a real fire occurs. At the University of Maryland, Baltimore, we had four fires in four separate high-rise buildings programmed for three-floor evacuation during 2006. During each fire, occupants were confused when alarms did not sound on their floors and yet they realized there was a fire in their building because of smoke odors and fire apparatus outside. One such fire was in a fourteen-story combination office and parking garage complex. Three floors of offices are atop the parking garage. The alarm went off on the seventh, eighth, and ninth floors of the parking garage. A vehicle fire occurred on the eighth floor (figure 2.31). A single sprinkler activated on the eighth floor and controlled the vehicle fire until the fire department arrived. The parking garage is an open-style garage, so smoke from the vehicle fire drifted outside and up past the windows of the office area. This created a great deal of confusion and resulted in many people self-evacuating. This action placed them in greater danger than if they had stayed on their floors. Firefighters were working in one of the stairwells and had the stairwell door open. The stairs and landing became wet from the water and some smoke entered the stairwell. Those evacuating and unnecessarily using the stairwell were exposed to smoke and slip hazards from the water in the stairwell. Getting this particular building sprinkled during design became a very difficult task. The university fire marshal required that the entire building be sprinkled. The contractor appealed the sprinkler requirements to the state fire marshal and the state fire prevention commission, who both upheld the sprinkler requirement. Experts testified on behalf of the contractor, stating that

FIGURE 2.31 A fire occurred in a fourteen-story combination office and parking garage complex. Three-floor evacuation was used in this building. When the fire occurred, it caused a great deal of confusion in the occupied floors of the building that were not evacuated.

FIGURE 2.32 An oxygen/acetylene torch was being used and the acetylene cylinder caught fire, filling the entire twelfth floor of a fourteen-story high-rise with smoke. Once again, three-floor evacuation was used and created confusion on the other floors of the building that were not evacuated.

a sprinkler would not put out a fire in a car in an open garage anyway because the heat would not build enough to cause the sprinklers to activate. This fire proved them wrong. A single sprinkler head not only activated, but it controlled the fire until the fire department arrived and kept cars on either side of the fire vehicle from experiencing any fire damage.

Several months later, another fire occurred in the same building when construction workers were welding on the twelfth floor, which was being renovated. An oxygen/acetylene welding torch was being used and the acetylene cylinder caught fire, filling the entire twelfth floor with smoke (figure 2.32). While the two occupied floors above the fire floor were evacuated, because of the three-floor evacuation, the parking garage was not. This caused confusion with people on the parking garage floors when the fire started. No alarms were sounding and yet they were not allowed to get to their cars to remove them or park any additional cars in the building. Parking attendants in all university parking garages are trained as fire wardens and when the fire alarm sounds they are supposed to stop vehicular and pedestrian traffic from entering a parking garage until the fire department gives the all clear. The parking garage was eventually closed during the fire even though the fire alarm was not sounding on the garage floors.

A similar situation occurred when the roof of an eight-story building caught fire (figure 2.33). No alarms went off at all in the building because sprinklers did not cover the roof and no detectors were present to activate the fire alarm. The building alarm was not activated until the fire department arrived, and then only the two floors below the penthouse. People saw two alarms of fire equipment on the street and experienced a burning smell in the building and many began self-evacuation.

FIGURE 2.33 Confusion over three-floor evacuation occurred again when the roof of an eight-story building caught fire.

Recently, yet another fire occurred on the fifth floor of a fourteen-story research/classroom building (figure 2.34). Elevator shafts carried smoke and burning smells through out the building even though the fire alarm was only activated on the fourth, fifth, and sixth floors of the building. Once again there was confusion throughout the building. Authorities having jurisdiction need to give careful consideration to the concept of using three-floor evacuation for reducing building disruption during false

FIGURE 2.34 Recently, another fire occurred on the fifth floor of a fourteen-story research/classroom building, resulting in confusion over the three-floor evacuation once again.

FIGURE 2.35 Voice-type fire alarm devices also allow for emergency public address (PA) system capability from the fire command center to the entire building, individual floors, elevator cars, and stairwells.

alarms compared to the confusion that may occur during a fire before they decide to use three-floor evacuation. While it sounds like a good idea in theory, three-floor evacuation does not always work real well during an actual fire. The University of Maryland has decided to discontinue the use of three-floor evacuation.

Voice-type fire alarm devices also allow for emergency public address (PA) system capability from the fire command center to the entire building, individual floors, elevator cars, and stairwells (figure 2.35). Using the manual PA system overrides any messages being transmitted over the fire alarm system, allowing firefighters or other authorities to give special instructions as needed during evacuations and other firefighting activities. If the fire alarm system is programmed for three-floor evacuation, the PA system could be used to assure occupants of other floors that they do not need to evacuate and update them with the progress of the fire in the building. However, this may cause confusion and unnecessary evacuations as well. It can also be used to notify persons in areas of refuge or rescue assistance to return to their floors or that the fire department is coming to evacuate them. PA system capability can also be used to evacuate individual floors during chemical releases, gas leaks, and other emergencies or if additional floor evacuations are required beyond the initial automatic selective evacuation.

Firefighter two-way communication phones are required for high-rise buildings when the fire department wants them installed (figure 2.36). At first glance you might think that you do not need them because you have radios. However, radios do not always work inside buildings or at all locations in buildings. Having a backup communications system in the building might prove very useful under some circumstances. For example, during bomb threats or suspicious package incidents, radios should not be used; however, the firefighter phones could provide communications in their place. Firefighter phones also provide the capability of secure communications not afforded by fire department radios. Firefighter phones allow for communication

FIGURE 2.36 Firefighter two-way communication phones are required for high-rise buildings when the fire department wants them installed.

between firefighters on each floor and the fire command center. Up to five handsets can be in operation at one time, allowing for limited conference calling. Phone jacks are generally installed at each floor landing in each stairwell, in the elevator lobbies, elevator cars, at fire pumps, and sometimes in electric transformer rooms and elevator machine rooms. The AHJ may require them in other locations. Phone handsets are stored in marked cabinets in the fire command center. Types of phones are also available that have the phone at the location required instead of just a phone jack. These units have additional cost associated with them but can be more convenient because you do not have to issue and carry handsets through the building. Controls for the voice fire alarm system and firefighter phones are required in the fire command center. These are often incorporated as part of the fire alarm control panel depending on the manufacturer (figure 2.37). They can also be in separate control panels. When a high-rise building is to be constructed and firefighter phone systems installed, make sure exact locations of phones are marked on the control panel. Systems can be designed and marked to show the actual location of a firefighter phone jack, so if a firefighter becomes lost and does not know where he is, plugging the handset into the phone jack would identify the location of the firefighter to personnel in the fire command center. Firefighter phones can also be used as part of the communications system for persons in areas of rescue assistance. At the University of Maryland, Baltimore, we assign firefighter phone handsets to persons confined to wheelchairs who cannot evacuate the building when the fire alarm is activated. They go to the area of rescue assistance and plug in the firefighter phone and can let firefighters and others know where they are and be given instructions. Being able to talk to someone who knows the conditions in the building helps to make them feel more secure. Fire command centers for high-rise buildings, when present, must incorporate the following controls and systems for firefighter use:

FIGURE 2.37 Controls for the voice fire alarm system and firefighter phones are required in the fire command center. These are often incorporated as part of the fire alarm control panel, depending on the manufacturer.

- Voice fire alarm system panels and controls
- Fire department two-way telephone communication service panels and controls (firefighter phones)
- Firefighter phone system handsets (twelve to fifteen minimum)
- Fire detection and fire alarm system annunciation panels
- Elevator floor location and operation annunciators
- Sprinkler valve and water flow annunciators
- Emergency generator status indicators
- Controls for any automatic stairway door unlocking system
- Fire pump status indicators
- Controls for smoke evacuation systems
- Controls for stairway pressurization
- A telephone for fire department use with controlled access to the public telephone system
- Any other fire protection system controls (such as movable fire barriers)

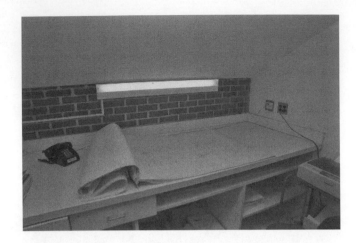

FIGURE 2.38 It is also helpful if as-built construction plans are located in the fire command center for fire department use during emergencies.

It is also helpful if as-built construction plans are located in the fire command center for fire department use during emergencies (figure 2.38). Trying to locate building plans during an emergency is not always an easy task, so if they are already in the fire command center they can prove very beneficial. Knox® Boxes with keys for building doors, elevators, and fire protection systems can be located in the fire command center. Additional building information such as locations of elevators and areas of refuge or rescue assistance can help firefighters unfamiliar with a building to locate these areas. Computers with Internet access can be a helpful resource in the fire command center as well.

Once the location of the fire alarm device causing the alarm has been determined, response personnel need to investigate the location to see whether a fire has occurred or another type of emergency caused the alarm or whether it was a false alarm. The use of elevators during fire incidents is very controversial. Elevators are not usually used for occupant evacuation. One exception is trained operators evacuating occupants with special needs. They should, however, be designed for the fire service use. Firefighters should be familiar with elevator emergency operation controls because they may want to use them when searching for the fire alarm device(s) that may be in alarm or the actual location of smoke and fire in the building. Elevators may also be used by firefighters to evacuate persons in areas of rescue assistance. Firefighters may also choose to use stairs to investigate the cause of a fire alarm.

NFPA 1, NFPA 101, NFPA 5000, the IBC, and the IFC all require that identification signage be provided inside stairwells at every level (figure 2.39). These standards all require stairwell signs in buildings over a certain height, but the height thresholds vary. NFPA 101, the *Life Safety Code*, requires stairwell signs in new buildings over three stories or existing buildings over five stories. Signage should show the stair identifier, floor level, terminus of the top and bottom, roof accessibility, discharge level, and direction to exit discharge. On floors that require

FIGURE 2.39 Firefighters may choose to use the stairs to investigate the cause of the alarm. NFPA 1, NFPA 101, NFPA 5000, the IBC, and the IFC all require that identification signage be provided inside stairwells at every level.

upward travel to reach the exit, a directional indicator should also be provided. It is important that these signs be located five feet above the floor and be visible with the stairway door open or closed. In hotels or other buildings with room or suite numbers, the signs should also include the room or suite numbers most directly accessed by each stair on every level, (e.g., second floor of stairway three has direct access to rooms 202 through 256). The latter signage would be extremely important when certain stairways provide no access to some sections of the building. Buildings more than three stories in height above grade should have roof access. The IBC and IFC require this, except for buildings with steeply pitched roofs (with a more than 4:12 slope). As stated above, the IFC, the IBC, NFPA 5000, and NFPA 241 contain special construction/demolition requirements. One stairway should be completed as construction advances. Conversely, as demolition progresses, one stairway should be maintained. These standards also address lighting and fire rating of the enclosure.

Building and fire codes typically require that stairs be large enough to accommodate exiting occupants. Fire service personnel who may use the stairs are not factored into exit capacity calculations. In situations where occupants are still exiting and firefighters are using the same stairs to enter the building ("counterflow"), the evacuation may take longer (e.g., World Trade Center, September 11, 2001). Furthermore, in most cases, stairway capacity is designed based on the floor with the highest occupant load. Typically, stairs are not widened as one travels in the direction of egress unless the stairs converge from both above and below. This approach assumes that people will evacuate in a phased manner, beginning with the floor(s) closest to the fire origin. In an immediate general evacuation or when people from other areas self-evacuate, the increased load will slow evacuation. Both of these bottlenecks will be made worse as the height of the building increases. Additionally, total evacuation is becoming more commonplace due to concerns about terrorism. An effective solution to the counterflow issue is a dedicated firefighting stairway. Codes in the United Kingdom contain specifications for such firefighting stairs, elevators, and intervening lobbies in buildings of certain height. Current U.S. codes do not require dedicated stairways or elevators for firefighters. The disadvantages of dedicated firefighting stairways include: cost, space, and the effort needed to keep them clear and in operating order. A solution to egress delays caused by either counterflow or total evacuation is to provide additional stairs or widened stairs. Cost and space are also disadvantages of this solution. These issues currently remain unsolved in the code community; however, a designer may encounter these issues on projects for large high-security or high-profile facilities. Further guidance on the movement of people in buildings can be found in the Society of Fire Protection Engineers' publication *Human Behavior in Fire*.

While investigating a building to locate alarm devices that have activated, keep in mind that some smoke detectors and heat detectors have blinking pilot lights and, when in alarm, display a solid light (figure 2.40). Others do not light unless in alarm. Some have one color, such as green, under normal operations and show solid red when in alarm. Others may just show blinking red when in alarm. It may be necessary to view other devices in the building to determine what type of signal the device displays under normal conditions. Many things can cause detection devices to alarm other than fire or smoke. If no smoke or fire is present, then it was likely a false alarm. If a manual pull station has been activated, this station must be reset before the fire alarm system can be reset. Fire alarm systems will not reset until all devices in the system are reset or out of alarm. Manual pull stations are the source of many false alarms. Several commercial devices are available that make it more difficult to falsely activate a manual pull station and provide protection against accidental bumps. They also can keep children from activating the devices. These devices have a plastic cover that must be opened before the pull station can be activated.

FIGURE 2.40 While investigating a building to locate alarm devices that have activated, keep in mind that some smoke detectors and heat detectors have blinking pilot lights and, when in alarm, display a solid light. Others change from green to red.

Some also have local alarms that deter mischief alarms. During plans review you may want to require the use of manual pull station covers to help reduce false alarms.

PREPARING TO DEPART THE SCENE

It is important that firefighters make sure that fire alarm systems are placed back in service before they leave the scene or follow up through the fire prevention bureau to make sure that those systems are reset in a timely manner. Because of the technical nature of fire alarm and fire protection systems and because the cause of alarm cannot be determined after the system is reset, firefighters should not reset systems. Resetting should only be performed by trained fire alarm system technicians. All fire responses should be documented through detailed reports. This is very important in the case of false alarms because documentation can identify locations where repeated alarms occur and the cause. If facilities incur repeated false alarms, the fire prevention bureau should investigate to determine whether the system is installed properly or there are mechanical defects. False alarms should not just be passed off as necessary nuisances of running a fire department. Repeated false alarms not only create complacency among firefighters responding but also among building occupants, who may not evacuate when there are too many false alarms in a building. For example, at the Charles Towers Apartments in Baltimore, a fire occurred on one of the upper floors. There had been so many false alarms in the building that when a real fire occurred, many people ignored the fire alarm when they could have gotten out safely on their own. The building was not sprinkled and firefighters had to be lowered onto the roof from a helicopter to rescue people who should have evacuated on their own.

3 Fire Alarm Systems

INTRODUCTION TO FIRE ALARM SYSTEMS

Certain types of occupancies are required to have automatic or manual fire alarm systems by the National Fire Protection Association (NFPA) *Life Safety Code* and model building and fire prevention codes. Fire alarm systems may also be installed even if they are not required by any code. All that is required is a good job of salesmanship by the authority having jurisdiction (AHJ). The primary purpose of a fire alarm system in an occupied building is to alert occupants to a fire and start a timely evacuation of the building. Fire alarm systems can also be used to evacuate a building for other types of emergencies that are not fires, such as chemical releases and gas leaks. A fire alarm system consists of interconnected automatic smoke, heat and fire monitoring devices, manual activation devices, notification devices, and auxiliary controls. The system is designed to alert building occupants to fire or dangerous conditions and provide emergency responders with information about those conditions. Clear and concise information will enable responders to operate efficiently and safely at an incident scene. Fire alarm systems monitor alarm-initiating devices such as manual pull stations; automatic smoke, heat, and fire detectors; or water flow indicators. Fire alarms also interface and control elevators and building HVAC systems as well as smoke control and smoke evacuation systems. Fire alarm systems operate in the following manner: When a signal is received, the control components process it via software programs or relays. The system then activates audible and visual evacuation notification devices; sends a remote signal to the fire service or other authorities; displays the location of the alarm; recalls elevators; shuts fire doors on magnetic hold-open devices; initiates smoke control systems; starts stair pressurization; and controls ventilation systems. Systems can vary widely in complexity. A basic, fundamental system consists of a control panel, initiating devices, and notification devices. On the other end of the spectrum are complex selective voice evacuation systems with integrated fire department phone communications systems. Detection systems have devices that automatically sense fire or its by-products. Detection systems are often integrated into fire alarm systems, and both will be covered in this chapter.

Building and fire codes often specify requirements for fire alarm systems. Commonly used codes include the IBC, *Uniform Fire Code*, NFPA 1, NFPA 5000, and NFPA 101. The *National Fire Alarm Code*, NFPA 72, is a comprehensive installation standard. This code, along with the fire alarm wiring portion of the *National Electrical Code*, NFPA 70, sets the requirements for design, installation, and maintenance. In addition, OSHA standards create requirements with respect to employee alarm systems. This chapter covers fire service personnel interaction with fire alarm systems and provides guidance for designers to facilitate operational

FIGURE 3.1 Auxiliary signals are received utilizing the same equipment and methods as alarms that come from manually operated municipal fire alarm boxes located on street corners.

efficiency. Elevator control, often interconnected with the fire alarm system, is discussed in chapter 7. There are several types of fire alarm systems identified in terms of who monitors signals and how the fire department is notified. We will talk here about the most common types including auxiliary, central station, municipal, proprietary, remote supervising station, and protected premises.

Auxiliary Fire Alarm System

Auxiliary fire alarm systems are connected to the municipal fire alarm system and transmit a signal to the public fire service communications center. These signals are received utilizing the same equipment and methods as alarms that come from manually operated municipal fire alarm boxes located on street corners (figure 3.1). Permission from the AHJ has to be received in order to transmit signals over the municipal fire alarm system. Only fire alarm signals may be transmitted over a municipal or auxiliary fire alarm system.

Central Station Fire Alarm System

Central station fire alarm systems are designed to receive signals from a protected premises at a constantly attended location operated by a private company that provides central station service. Such companies must be listed by either Underwriters Laboratory (UL) or Factory Mutual Global Research (FM). When a signal is received at a central station facility, they must notify the public fire service communications center and designated persons at the protected premises or at their homes. A service technician must be dispatched to the protected premises within one hour of the receipt of the alarm to reset the system where required. Supervisory and trouble signals must also be investigated by a service technician. Supervisory signals must be investigated within one hour and trouble signals within four hours.

Proprietary Fire Alarm Systems

Signals received from a proprietary fire alarm system are monitored at a constantly attended supervising station located either on the premises or at another location of the property owner. Proprietary systems transmit alarm, supervisory, and trouble signals to the supervising station. Upon receipt of an alarm, supervisory, or trouble signal, a person must be sent to investigate. Fire alarm signals must also be reported to the public fire department.

Remote Supervising Station Fire Alarm System

Remote station fire alarm systems transmit alarm information to a remote location, which may include the public fire service communication center, fire station, or fire alarm monitoring company. Signals can be transmitted over data lines, fiber-optic lines, microwave and other types of radio systems, and public telephone lines. There must be a primary means of transmission and a backup secondary means. Supervisory and trouble signals may also be transmitted to the remote station.

Protected Premises Fire Alarm System

A protected premises fire alarm system is designed to notify building occupants that they must evacuate the protected building. This type of fire alarm system does not automatically notify the fire department. Someone at the protected premises must manually contact the fire department communications center. If a protected premises is unoccupied when a fire alarm is activated, fire department notification depends on someone outside the building hearing the alarm and calling the fire service communications center.

Municipal Fire Alarm Systems

Municipal fire alarm systems are designed to send signals from remote fire alarm boxes (figure 3.1) on city streets directly to the public fire service communications center. Specific information about the installation, operation, maintenance, and testing of these fire alarm systems can be found in NFPA 72, the *National Fire Alarm Code*. These types of systems used to be in widespread use across the country, but their use has been greatly reduced over the past few decades. Some of the automatic street-corner boxes have been replaced by emergency phones connected directly to the public fire service communications center.

ANNUNCIATOR PANEL AND SYSTEM DEVICES

The first thing firefighters responding to a fire alarm in a building should look for upon entering the facility is the fire alarm annunciator panel (figure 3.2). Annunciator panels contain information about the condition of the fire alarm system in the building, location of the alarm, and type of device in alarm (figure 3.3). Annunciator panels are a mirror image of the information available at the fire alarm panel, but the fire alarm panel is usually located in a mechanical room or electrical room somewhere in the building basement or penthouse. They may also be located at the building entrance in place of

FIGURE 3.2 The first thing firefighters responding to a fire alarm in a building should look for upon entering the facility is the fire alarm annunciator panel.

the annunciator or other location within the building. It is usually not convenient to get information from the fire alarm panel quickly because of its location. Therefore, the location of the annunciator panel is critical. If the building is a high-rise, the fire alarm panel will be located in the fire command center (FCC) along with the annunciator or

FIGURE 3.3 Annunciator panels contain information about the condition of the fire alarm system in the building, location of the alarm, and type of device in alarm.

FIGURE 3.4 If a building is a high-rise, the fire alarm panel will be located in the fire command center (FCC) along with the annunciator or duplicate annunciator.

duplicate annunciator (figure 3.4). Therefore, the annunciator is the most important fire alarm system component in terms of immediate information available for responders. There are often unique circumstances concerning locations, types, and existence of annunciator panels. Generally, if there is one, it will be just inside the building in the lobby or entryway. High-rise buildings will have an FCC and the annunciator will be in the FCC or may also be duplicated at the main entrance if the FCC is not located near the entrance. Each building should have its own annunciator, even if a single fire alarm control system serves multiple buildings. Fire service operations would be delayed if it were necessary for one unit to report to a given building to check the annunciator and then relocate (or direct another unit) to investigate origination of the alarm. In large complexes, an additional master annunciator could assist the fire department in locating the building where an alarm originates much more quickly.

Older fire alarm systems may not be very sophisticated; newer systems should have floor plans of the building floors with LED lights showing locations of devices in alarm (figure 3.5). Some of the information on the annunciator is dictated by code. The code that deals with most aspects of fire alarm installation is the National Fire Protection Association (NFPA) Code 72, the *National Fire Alarm Code*. Additional guidance for the installation of fire alarm systems can be found in NFPA 70, the *National Electrical Code*. Local jurisdictions may have amended their codes or made more stringent requirements for fire alarm installations. Building codes, fire prevention codes, and the *Life Safety Code* may also have fire alarm requirements for certain occupancies and situations. There are also associated elevator code issues that require coordination with the elevator code ASME A17.1 (see chapter 7) when installing a fire alarm system or elevators (figure 3.6).

The sole purpose of an annunciator panel is to provide information to those responding to an alarm in a building, usually the fire department. The more information provided on the panel, the more help it will be to responders when the fire alarm

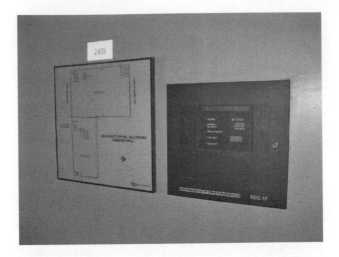

FIGURE 3.5 Older fire alarm systems may not be very sophisticated; newer systems should have floor plans of the building floors with LED lights showing locations of devices in alarm.

system goes into alarm. How much input does your fire prevention bureau provide as to the design of the annunciator? What features does your department want on the annunciator to assist them when responding to buildings? Have you given thought to the creation of a boiler plate annunciator so that every annunciator installed contains the same type of information? Most plans for new fire alarms systems are reviewed by the fire prevention bureau, fire marshal, building code official, or fire chief. This is an opportunity to provide input to the design of the annunciator panel. Unfortunately, many departments do not take this opportunity and just settle for what is proposed on

FIGURE 3.6 There are also associated elevator code issues that require coordination with the elevator code ASME A17.1 when installing a fire alarm system or elevators.

the plans. Remember, the annunciator generally is of no use to the building owner, occupants, or anyone else but the fire department, so why should it not be designed to our specifications? Sure, it will require more work initially by your department, but it will certainly save time when your department responds to an alarm in the building. It all depends on the focus of your organization. Plans review and inspection personnel should be educated extensively about fire alarm and fire protection systems. Firefighters should also have training covering the operation and code requirements for fire alarm and fire protection systems. They are the front line of the organization's fire prevention efforts, the ones the public sees most and the ones who have the most opportunity to interact with the public. When firefighters respond to a fire alarm, they should be prepared to help determine why the alarm activated. If the alarm turns out to be false—and most do—they should make recommendations on preventing future false alarms. Information about the fire alarm system should be passed on to the fire prevention bureau for follow-up. False alarms can be prevented with a little effort by firefighters, fire prevention personnel, and building owners.

Annunciator panels display alarm information in different ways. Some have lights or LEDs that are labeled with information about devices activated. Alphanumeric annunciators have a readout-type display that may be programmed to show very specific information describing the alarm signal, including the device that went into alarm and its location. Some fire alarm system installations have both LED annunciators and alphanumeric readouts on the annunciator, with additional alphanumeric readouts on the fire alarm panel as well. A printer is yet another means of annunciation (figure 3.7). Printers may be located in the fire command center of a high-rise building or near the fire alarm control panel in other types of buildings. Printers are usually provided with the fire alarm system in conjunction with other annunciation devices. In very simple systems, the control panel serves as the annunciator.

FIGURE 3.7 Printers may be located in the fire command center of a high-rise building or near the fire alarm control panel in other types of buildings.

In such cases, its location and features should meet all annunciator requirements. The annunciator panel may also store building plans and diagrams. These are then quickly accessible to firefighters. A note outside the panel can indicate that it contains building plans or diagrams. All annunciators should include floor, zone, and device annunciation. Local fire or building codes may dictate zone size. The annex of NFPA 72 specifies a maximum zone size of twenty thousand square feet and three hundred linear feet. The zone limitations in both the IBC and NFPA 5000 are 22,500 square feet and three hundred linear feet. Zone boundaries should coincide with fire ratings, smoke ratings, or building use boundaries. Zone descriptors, whether labels located next to lamps or alphanumeric displays, should provide pertinent information for fire service personnel. Designers should assume that users will not be familiar with the building. Descriptors should be intuitive and rapidly decipherable. As the building, layout, tenants, or room names change, building owners should update descriptions.

Flow switches or pressure switches indicate water flow (figure 3.8) from the sprinkler system in a building to the fire alarm control panel and the annunciator panel. To direct the fire department to the appropriate area where the sprinkler has been activated, it is important that the zone indication shows the area covered by the sprinkler system. The location of the switch itself is not important for fire department response operations. There is nothing the flow switch can do for the fire department except identify the location where the sprinkler system has activated, which may indicate that a fire has occurred. Fire alarm devices indicate a situation requiring emergency action and normally activate evacuation signals.

Alarm devices include:

- Manual pull stations
- Sprinkler flow
- Smoke detector

FIGURE 3.8 Flow switches or pressure switches indicate water flow from the sprinkler system in a building to the fire alarm control panel and the annunciator panel.

- Beam detector
- Heat detector
- Kitchen cooking equipment extinguishing system
- Clean agent systems
- Halon systems
- Flame detectors
- CO_2 industrial systems

Smoke and heat detectors should be further identified on the annunciator by mounting location:

- Area
- Under floor
- Air plenum
- Elevator lobby
- Elevator hoistway
- Stair tower
- Elevator pit
- Elevator machine room

Supervisory devices indicate abnormal conditions. They signal a need for nonemergency action, such as repair or maintenance, and they should not cause an evacuation alarm or notify the fire department.

Examples of supervisory devices include:

- Valve tamper switch
- Dry sprinkler high or low air pressure switch
- Pre-action sprinkler low air pressure switch
- Water tank low temperature or low water level indicator
- Valve room low air temperature indicator

Some jurisdictions require devices that are subject to unwanted alarms (primarily duct or air plenum smoke detectors) to be supervisory (figure 3.9). Codes also allow elevator machine room, elevator pit, and elevator shaft heat and smoke detectors to be placed on supervisory signal. Beam detectors may also be placed on supervisory if desired by the authority having jurisdiction. Some fire alarm devices control certain building features, such as fans, doors, or dampers. They may be shown as "alarm" or "supervisory," depending on the preference of the code official.

Examples of alarm or supervisory devices:

- Duct smoke detectors
- Air plenum smoke detectors
- Under floor detectors
- Door closure smoke detectors
- Elevator hoistway smoke detectors
- Elevator machine room smoke detectors

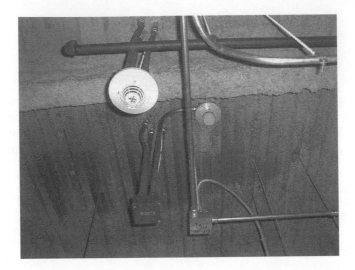

FIGURE 3.9 Codes allow elevator machine room smoke and heat and duct detectors to be placed on supervisory signal.

- Heat detectors for elevator shutdown (shunt-trip)
- Stair smoke detectors
- Beam detectors

Status indicators give information about whether the main fire alarm power is on or they report on the condition of devices external to the alarm system.

Examples of status indicators:

- Main system power on
- Main system trouble
- Fire pump running
- Fire pump fault
- Fire pump phase reversal
- Generator run
- Generator fault
- Stair doors unlocked
- Smoke control system in operation

Controls are switches that operate features external to the fire alarm system. Examples of control switches:

- Remote pump start
- Remote generator start
- Smoke control manual switches
- Stair pressurization manual switches
- Stair unlocking switches
- HVAC switches

If an annunciator shows any location-related information that is not obvious, a graphic diagram should be provided. Examples are zone boundaries, room names, or room numbers. Diagrams enable firefighters to determine where to investigate alarms originating in locations with designations such as "Zone 2 East," "Suite 121," or "Main Electric Room." The graphic display may be a separate, printed diagram mounted on the wall adjacent to the annunciator. Or it may be integrated with the annunciator, in which case it is called a "graphic annunciator." Some jurisdictions may require graphic annunciators. The design of the diagram is very important in enabling firefighters to rapidly obtain needed information. Fire departments may have regulations or policies outlining their requirements for preferences. Some code officials require annunciators throughout their jurisdiction to have standardized features. Orientation of the diagram will be important in aiding firefighters to visually process the information it contains. The farthest point of the building beyond the annunciator's location should be at the top of the diagram. Designers should begin with the building's outline in creating diagrams. Zones would be identified by the boundary lines between them. Likewise, for alarms designated by room, suite, or tenant, these locations should be shown. A "You Are Here" indicator shows the viewers where they are in the building. Floor plans should also show stair identification numbers, elevator locations, fire pump room location, and other information that will help firefighters better navigate the building during an emergency.

NFPA 13 permits most sprinkler zones to cover as much as fifty-two thousand square feet. Therefore, multiple alarm zones may cover one sprinkler zone. If there is one sprinkler zone on a floor and multiple alarm zones, lamp or LED annunciators should report only the floor and device type. An alarm from another device type will light the appropriate zone lamp. If there are multiple sprinkler zones per floor, and sprinkler and alarm zone boundaries are not coordinated, separate diagrams can show each. Consistent designations for any floor indications used in the building will avoid confusion. For example, it is imperative that floor designations on the signs mounted in stairways, elevator cars, and elevator lobbies be consistent with the annunciator so the firefighters report to the correct floor (figure 3.10). In addition to information about floors, zones, and devices, many features of the building could be shown on the diagram. These include fire protection systems and building components that the fire department needs to be aware of. Designers should remember that modifications to the building or its layout may require changes to the diagram. An annunciator with inaccurate information could be worse that no annunciator at all.

Fire Alarm Panels

Fire alarm panels are the control center for the fire alarm system (figure 3.11). Alarm systems can be addressable or non-addressable. Addressable fire alarm systems indicate where an alarm has originated, what type of device, and what type of signal has been transmitted. Non-addressable systems are usually older systems and only indicate a floor or zone where the alarm has originated but not the type or location of the initiating device. Addressable systems are much more advantageous than non-addressable systems. It is important that responding firefighters know not only on which floor the alarm has occurred but what type of device has been activated and its location, and they need to get that information as quickly as possible.

FIGURE 3.10 It is imperative that floor designations on the signs mounted in stairways, elevator cars, and elevator lobbies be consistent with the annunciator so that firefighters report to the correct floor.

FIGURE 3.11 Fire alarm panels are the control center for the fire alarm system.

Otherwise, it will require a search of the entire floor to locate the device that has caused the alarm to activate. Once the cause of a fire alarm activation is determined and the building has been declared safe, the alarm must be cleared from the system at the fire alarm panel. Before a panel can be reset, the device in alarm must be cleared first. Pull stations will require a manual reset, which may involve a key or other tool for access to the reset button. Smoke and heat detectors must clear themselves before the panel will reset. The devices must be clear of smoke or heat before they will reset. Panels also contain all of the connections for the wiring of the system from throughout the building. Any repairs to the fire alarm system originate with a visit to the alarm panel. Panels are generally located in mechanical areas or electrical rooms within protected buildings. In high-rise construction they are located in the fire command center as discussed in chapter 2. To enhance survivability of the fire alarm system, codes require that a smoke detector be placed near the fire alarm control panel (figure 3.12). When a new fire alarm system is installed in a renovated

FIGURE 3.12 To enhance survivability of the fire alarm system, codes require that a smoke detector be placed near the fire alarm control panel.

or new building, panels are generally placed in one hour rated rooms to help ensure survivability of the fire alarm equipment. However, existing systems may not have the panel installed in a rated room. A licensed electrician with the guidance of a fire alarm company usually installs a fire alarm system. State fire marshals in many states and some local jurisdictions license fire alarm installers and designers. This process helps ensure that persons designing and installing fire alarm systems will install them properly. NFPA 70, the *National Electrical Code*, and NFPA 72, the *National Fire Alarm Code*, are documents used for guidance with proper system installation and testing.

Technology developed in the seventies and eighties brought about the use of smoke detection and alarm devices to wake sleeping people and alert them to a fire (figure 3.13). This technology was implemented to reduce the large loss of life that was occurring in residential fires. Smoke alarms have been responsible for cutting the fire death rate in residential fires almost in half since their inception. Smoke detectors have been incorporated into commercial fire alarm systems by building, fire, and elevator codes to accomplish specific tasks in the overall picture of life safety within occupancies other than residential. In some cases, smoke detectors have been used with other life safety components to gain equivalency to sprinkler system requirements in hospitals, nursing homes, educational and other occupancies. While I do not agree with this practice, it is nonetheless something that is allowed by some AHJs. If a building needs to have sprinklers, then the owners should provide a plan of action to install sprinklers. I would hate to have to explain to someone's relatives that their loved ones died because we did not want to create a financial burden for the owners of a building by requiring the installation of a sprinkler system.

Building or fire codes usually require fire alarm systems to automatically alert the responsible fire brigade, fire department, or other emergency response forces.

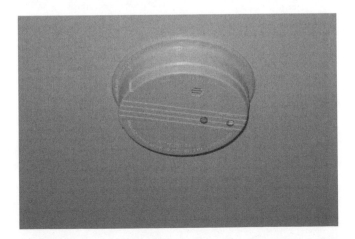

FIGURE 3.13 Technology developed into the seventies and eighties, which brought about the use of smoke alarm devices to wake sleeping people and alert them to a fire.

The important consideration for fire department response is reporting the correct location. Often an alarm service or off-site location will receive the alarm signal and then retransmit it to the fire department and/or fire brigade. It is crucial that the address reported to the fire department match the address where the alarm originated. If a building has multiple addresses, the one with the fire alarm annunciator or fire command center should be reported. If a building includes separate, independent annunciators, coordinate the remote signal with the correct annunciator location. Larger buildings with multiple sections or multiple entrances can be confusing. If possible, remote fire department notification should include information on the section, wing, or entrance where units should report so that firefighters may investigate an alarm originating from the corresponding area. In addition, strobe lights at entrances corresponding to the alarm location for on-site notification can greatly assist the fire department.

MANUAL PULL STATIONS

Manual fire alarm pull stations are devices installed as part of the fire alarm system to allow occupants who discover a fire before an automatic device is activated to sound the alarm within the building manually and, if designed to do so, notify the fire department (figure 3.14). Some fire alarm systems are installed without any automatic devices and the only way the fire alarm system can be activated is by pulling a manual fire alarm pull station. This type of fire alarm system is referred to as a "manual fire alarm system." Manual fire alarm pull stations come in many sizes and shapes, depending on the manufacturer. Instructions for operation are generally located on the pull station handle. Manual fire alarm pull stations can also be utilized to perform other functions such as releasing exterior exit doors that are

FIGURE 3.14 Manual fire alarm pull stations are devices installed as part of the fire alarm system to allow occupants who discover a fire before an automatic device is activated to sound the alarm within the building and, if designed to do so, notify the fire department.

locked in urban areas for security. Analog pull stations are located next to exterior exit doors. When the fire alarm sounds, the doors are automatically released. If the power fails in the building, the doors are automatically released. Activation of the pull station next to the door also releases the door. This type of setup does not meet code but can be authorized by the local AHJ.

FALSE ALARM PROTECTIVE COVERS

Manual fire alarm pull stations have often been an attractive target for those who think they are playing practical jokes; to school children wanting an early dismissal; to children just not being properly supervised; to the criminal wanting a quick way out of a building or to create a distraction. If false alarms are to be reduced within a building, it is necessary to provide protective covers on manual pull stations as a beginning (figure 3.15). Putting protective covers on manual pull stations in a building is best and most economically accomplished during the construction of

FIGURE 3.15 If false alarms are to be reduced within a building, it is necessary to provide protective covers on manual pull stations as a beginning.

the building. When plans are reviewed, reviewers should make recommendations for the installation of alarmed protective covers over fire alarm pull stations to help prevent false fire alarms. Generally, they work by having a cover to open to get to the manual pull station. There are two basic types of covers, those with alarms and those without. Covers with alarms sound a local alarm when the cover is removed. A loud local alarm activates to startle the person removing the cover, which is designed to deter them from pulling the manual fire alarm pull station. If a real fire occurs, pulling the cover off is relatively quick, and the manual pull station can be activated to sound the alarm.

SMOKE ALARMS

Smoke alarms are generally used in residential occupancies or locations where people are sleeping (figure 3.16). Smoke alarms are single or multiple station alarms that respond to the presence of smoke and warn occupants that a fire may be occurring. Each smoke alarm is equipped with a detection device and an audible alarm. Some smoke alarms come equipped with visual devices as well and some with an emergency light to light the way when it is dark at night. The detection and alarm function of the devices are self-contained within each alarm. Smoke alarms may be battery operated, hard-wired to the building electric system, or a combination of hard-wire with battery backup. They should always be installed according to the manufacturer's recommendations. Generally, smoke alarms can be installed on the ceiling or on the wall. If installed on the wall, it is important to note that there is a dead air space where the wall meets the ceiling. Smoke alarms installed on the wall or ceiling should avoid the areas six inches either side of the wall ceiling junction. At a minimum, smoke alarms should be installed outside

FIGURE 3.16 Smoke alarms are generally used in residential occupancies or locations where people are sleeping.

FIGURE 3.17 Wireless systems are also available that have associated voice announcements that can be plugged into wall outlets in bedrooms.

sleeping areas of the home. Additional detectors should be installed on each level. The ideal situation is that smoke alarms are installed in every room on every level. Smoke alarms, however, should not be placed in areas where they are subject to false alarm. Kitchens and bathrooms should not have smoke alarms. Technology is currently available in which smoke alarms can be purchased that are interconnected either by hard-wire or radio signal to sound an alarm on all smoke alarms when one individual alarm senses smoke. There are also wireless systems available that have associated voice announcements that can be plugged into wall outlets in bedrooms (figure 3.17). In some cases, voice alarms work better for waking up certain age groups. While smoke alarms have a proven record of saving lives when fires break out, one of the major failing points of smoke alarms is missing batteries or failure to change batteries when needed (figure 3.18). Smoke alarms are usually equipped with an audible low-battery alarm. When this audible alarm is heard, the battery should be changed immediately. Firefighters and EMS personnel who respond to fires and other emergencies in residential occupancies should take the time to make sure smoke detectors are working, have batteries, and are properly installed (both numbers and placement). Do not be afraid to sell fire safety, smoke alarms and detectors, and sprinkler systems anytime you have the opportunity.

It is recommended that smoke alarms be tested monthly utilizing the test button on the device. Batteries should be replaced annually on an important holiday, birthday, or when time change occurs to or from daylight saving time. Energizer brand batteries teamed up with the International Association of Fire Chiefs in an annual campaign to remind people to "Change Your Clock/Change Your Battery." This program also provides free batteries to local fire departments to give away to people in their communities to replace smoke alarm batteries. Additional information can be obtained at the following Website: http://www.energizer.com/learning/FireSafety.asp. Generally,

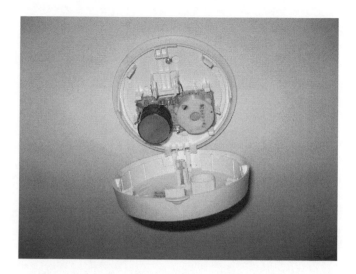

FIGURE 3.18 While smoke alarms have a proven record of saving lives when fires break out, one of the major failing points of smoke alarms is missing batteries or failure to change batteries when needed.

it is believed that smoke alarms have a useful life of ten years. Smoke alarms that have been installed ten years or longer should be replaced with new alarms and batteries. There are smoke alarms available that are equipped with ten-year lithium batteries, which will reduce the battery changing needs of the alarm. Ten-year lithium batteries can also be purchased separately to place in existing smoke alarms.

Smoke alarm technology may be either ionization type or photoelectric. Details of each type will be described below under the section on smoke detectors. Studies conducted on various age groups have shown that not all people will react to the audible sound of a smoke alarm. Younger children especially do not seem to respond well to the sound of the smoke alarm. Recent studies report that children are more likely to respond to commands from their mother's voice than the sound from a smoke alarm. KidsSmart Company has developed a residential smoke alarm that allows you to record the voice message that will be heard when the smoke alarm is activated (figure 3.19). With this type of smoke alarm, mom's voice calling the child's name could be recorded onto the detector and give children a better chance of waking up when a fire occurs in the home. In addition to repeating your child's name, you can issue instructions about leaving the home in the event of a fire, reminding your child not to open a hot door or to cover his or her face if smoke is present. There is even a fire drill button on the vocal smoke alarm so that you and your family can practice exiting the home as you would in the event of a real fire. The smoke alarms are battery operated and provide an audible signal and a voice signal, and some models provide an emergency light for maximum effect.

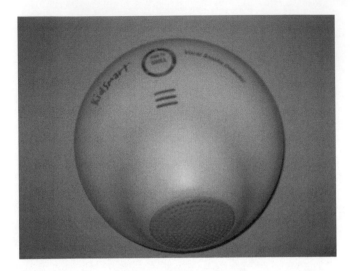

FIGURE 3.19 KidsSmart Company has developed a residential smoke alarm that allows you to record a voice message that will be heard when the smoke alarm is activated.

SMOKE DETECTORS

Smoke detectors, unlike smoke alarms, only sense the presence of visible or invisible particles of combustion; they have no audible signal. Unless they are part of a fire alarm system that has alarm devices, they by themselves do not sound alarms. Therefore, it is safe to say that smoke detectors will only be found in conjunction with occupancies that have automatic fire alarm systems installed. Smoke detectors can either be analog or non-analog as well as addressable and non-addressable. The concept of addressable and non-addressable is the same as for the fire alarm panel. When assembling the components for a fire alarm system, if you want an addressable system, you must also have addressable devices. Non-analog detectors are only capable of sending an on or off message to the fire alarm panel (normal or alarm condition). Analog detectors can provide alarm verification by sampling the air more than once to see if there is still a smoke condition with subsequent samplings. This can help to reduce the occurrence of false alarms. There are two primary types of smoke detectors, ionization and photoelectric.

Other smoke detection devices are also available and include beam detectors and duct detectors, which use the air sampling technique. In the case of duct detectors, a sample of air from the HVAC duct is pulled into a chamber where the smoke detector is located. If there is smoke present, the smoke detector will activate. Some new types of detectors use combination smoke and heat technology. They use both ionization and photoelectric technology in combination with heat sensors. Ionization smoke detectors make use of a radioactive source to ionize the air with two differently charged electrodes to sense the presence of smoke particles (figure 3.20). Ionization detectors are much less sensitive to materials that cause false alarms than are photoelectric detectors, but they also do not react as quickly to some types of smoke.

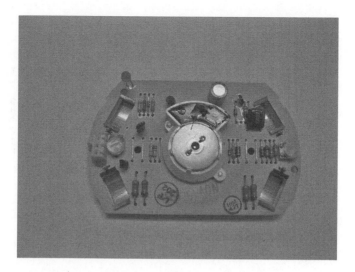

FIGURE 3.20 Ionization smoke detectors/alarms make use of a radioactive source to ionize the air with two differently charged electrodes to sense the presence of smoke particles.

Ionization detectors respond faster to flaming-type fires than do photoelectric detectors. Photoelectric detectors, on the other hand, respond quicker to smoldering-type fires (figure 3.21). Based upon this information, ionization detectors would seem to be better suited for occupancies where people are awake and would be aware of a fire taking place. Photoelectric detectors, on the other hand, would then be more

FIGURE 3.21 Photoelectric smoke detectors/alarms make use of a light source to sense the presence of smoke particles that enter the chamber in the center of the smoke alarm.

suitable for places where people would be asleep. Actually, a combination detector would really be the ideal type, since it would detect both types of fires regardless of the type of occupancy. There are two basic types of photoelectric detectors. They are light obscuration and light scattering. Light-obscuration detectors use a light source and a photosensitive sensor. When smoke particles enter the light path, obscuring the light reaching the sensor, an alarm condition is transmitted to the fire alarm control panel when it meets preset criteria. Light-scattering involves smoke particles entering a light path, resulting in the scattering of the light. The scattered light activates a photosensitive device, causing the detector to respond. Photoelectric detectors are much more prone to false alarms because dust, steam, bugs, and other foreign materials can cause the same light obscuration as does smoke.

Beam Detectors

Beam detectors are a special type of smoke detector (figure 3.22). Instead of using radioactive sources or electronic eyes to detect smoke, these detectors send a beam of light across a space to a light-sensitive source. When smoke passes through the beam of light and obscures the amount of light reaching the light-sensitive source, an alarm is initiated. Beam detectors are listed for the space they can effectively cover; for example, a given model may range from 33 feet to 330 feet (10 m to 100 m). Ranges will depend on the model and manufacturer. Make sure that the manufacturer's instructions are followed for installation to ensure proper operation of the detectors. Beam detectors are often used at the tops of atriums and other large, open spaces. My personal experience with this type of detector is that they are subject to false alarms if not properly installed and maintained or when something other

FIGURE 3.22 Beam detectors are a special type of smoke detector. Instead of using radioactive sources or electronic eyes to detect smoke, these detectors send a beam of light across a space to a light-sensitive source.

than smoke obscures the beam, such as steam, dust, and insects. In fully sprinkled buildings, beam detectors may be more of a problem than actually adding much of a life safety component to the building occupants. Where installations were made in buildings at the University of Maryland, we have made the beam detector transmit a supervisory signal rather than a general building alarm. The detector still activates the smoke evacuation system as designed but does not result in false building alarms. If the beam detector is sensing smoke from a fire, the sprinkler system will activate, triggering a flow switch, resulting in activation of the fire alarm, or someone will see smoke and activate a manual pull station. The smoke evacuation system will already be in operation from the beam detector. We do not believe this hinders life safety in any way but reduces the number of potential false alarms in the building.

DUCT DETECTORS

Duct detectors employ air sampling technology to extract a sample of air from a space (such as an air duct) and check it for smoke particles. This type of detector is commonly based upon photoelectric technology and is frequently used for air duct detection. Duct smoke detectors are mounted on the outside of air ducts and extract air samples to analyze from the duct air supply through a sampling tube and draw it into the smoke chamber on the outside of the duct (figure 3.23). If smoke is detected, the detector will go into alarm. One of the major problems with duct detectors is that dust, steam, and smoke from dust and other materials burning off of heating elements associated with ducts may cause detectors to false alarm. Codes allow for the authority having jurisdiction to permit duct detectors to be programmed to transmit a supervisory alarm rather than a general alarm. I would highly recommend that this approach be taken because it does significantly reduce false alarms while not greatly affecting life safety of building occupants. For example, a fire recently

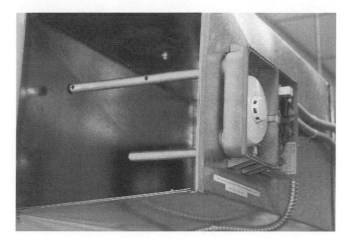

FIGURE 3.23 Duct smoke detectors are mounted on the outside of air ducts and extract air samples to analyze from the duct air supply through a sampling tube and draw it into the smoke chamber on the outside of the duct.

occurred in a duct system of a ten-story business occupancy in Baltimore. A fan motor burned out, melting the fan belt and grease in the motor. The duct detector sensed the smoke and shut down the unit as it was designed to do. Maintenance personnel discovered the burned-out motor when investigating the supervisory alarm from the building. At no time were any of the building occupants in danger. The real purpose of the duct detector is to shut down air-handling equipment when smoke is detected in the ducts, thus stopping the spread of smoke through the building's duct system. Since the air-handling equipment will be shut down when the detector operates, the detector has accomplished its mission. Once again, some other device in the building will detect the fire or someone will activate a manual pull station in an occupied building, setting off the fire alarm. Given the potential prevention of false alarms that may cause occupants to ignore fire alarms all together, I believe this is a good approach. This option should be weighed depending on the type of occupancy. Certainly occupancies where people are sleeping should be evaluated before making duct detectors supervisory. Remember, however, that places where people are sleeping should also have area smoke alarms outside sleeping areas. In a fully sprinkled building of any type of occupancy, the sprinkler system will likely control the fire and sound the alarm possibly before smoke reaches the ducts. Duct detectors sample air that is recirculated throughout a building. Some occupancies that contain chemical and biological laboratories operate with 100 percent outside air and have five to twelve air exchanges per hour (figure 3.24). There would be no advantage to duct detectors in this type of air-handling system because the air is not recirculated through the building. Most other types of occupancies, however, do recirculate air through the building to save on conditioning costs. Buildings with recirculated air should have detectors in their duct system as prescribed by the appropriate code.

FIGURE 3.24 Some occupancies, such as research facilities that contain chemical and biological laboratories, operate with 100 percent outside air and have five to twelve air exchanges per hour.

Duct detectors are required to be placed downstream of air filters and in front of branch connections in air supply systems having a capacity larger than 2000 ft³/min (944 L/s) according to NFPA. They are also required at each floor prior to the connection to a common return and prior to any recirculation of fresh air inlet connection in air return systems having a capacity greater than 15,000 ft³/min (7,080 L/s). The purpose of the duct detectors is to automatically stop their individual fans upon the detection of smoke resulting from a fire in the air-handling unit. Multifunction detectors combine smoke and heat detection in some models and heat and both types of smoke detection in other models. Models with heat, ionization, and photoelectric smoke are the best choice in terms of the most versatile type of detection device to ensure quick activation no matter what type of smoke or fire conditions are present.

CARBON MONOXIDE DETECTORS

Smoke alarms have been commonplace in residential occupancies for many years. Recently, the development of carbon monoxide detectors has brought another technology into the home (figure 3.25). Carbon monoxide (CO) is a colorless, odorless, and tasteless, nonirritating gas, which makes it almost impossible to detect by the human senses. Carbon monoxide is very toxic and a concentration of as little as 0.04 percent (four hundred parts per million) in the air can be fatal. When carbon monoxide is inhaled, it takes the place of oxygen in hemoglobin, the red blood pigment that normally carries oxygen to all parts of the body. Because carbon monoxide binds to hemoglobin several hundred times more strongly than oxygen, its effects are cumulative and long-lasting, causing oxygen starvation throughout the body. This results in chemical asphyxiation and death. Carbon monoxide is a product of incomplete combustion of carbon-based products, particularly natural gas, in home appliances.

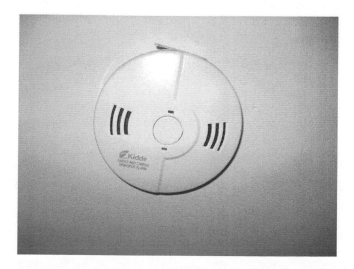

FIGURE 3.25 This is a combination smoke alarm/carbon monoxide detector located outside bedrooms in the hallway.

Additional hazards exist with carbon monoxide when gasoline-powered equipment is used indoors; for example, generators during power outages. Gasoline- or diesel-powered equipment should never be used inside of a building. Each year two to three hundred people unintentionally are injured or killed from carbon monoxide. Another hundred-plus die from CO but it is not known whether it was intentional (suicide) or unintentional (accidental). Other cases of CO death often go undetected for various reasons.

Many carbon monoxide–induced injuries and deaths could be prevented with the use of a residential carbon monoxide detector. These devices, similar in appearance to smoke detectors, can determine the presence of carbon monoxide gas and sound an alarm before it is too late for occupants to escape on their own. In fact, some companies sell a combination smoke alarm/carbon monoxide detector. Where a building has a fire alarm system installed, the fire alarm system can monitor CO detectors as long as the CO detector transmits a supervisory alarm that is identified as a carbon monoxide detector to the receiving location of the fire alarm system. Carbon monoxide detectors should be installed according to manufacturer's instructions. CO detectors should be installed outside sleeping areas. Additionally, if gas-supplied heating equipment is located in the basement, a detector should be installed at the top of the basement stairs. It would be helpful if CO detectors were hooked together to sound the alarm to the other locations. If you have a fireplace, gas heater, or other type of gas-fired heating equipment, a CO detector should be installed in the room where the equipment is located. CO detectors should also be installed in rooms where cooking equipment, refrigerators, and clothes dryers fueled by gas are located. Carbon monoxide buildup often occurs from defective heating equipment, insufficient makeup air, blocked or design-flawed chimneys, or the absence of required vents. Other sources of CO include barbecue grills, generators, other portable gasoline-powered equipment, and idling automobiles in attached garages. CO detectors have a useful life of approximately five years and should be replaced after this time. NFPA developed a standard for carbon monoxide detectors in 1993 titled NFPA 720 *Standard for Recommended Practice on the Installation of Carbon Monoxide Warning Equipment.*

HEAT DETECTORS

Although considered reliable and least prone to false alarm, heat detectors, when compared to smoke detectors, are the least sensitive and thus in some applications the least effective type of detection device. Great care needs to be taken in determining the appropriate use for heat-detection devices. Heat detectors were the first fire detection devices developed (figure 3.26). They were developed along with automatic sprinklers in the 1860s and continue in use today. Before and after the advent of smoke alarms, heat detectors were advocated as effective detection and alarm devices for residential occupancies. Although before smoke alarms were developed, heat detectors were better than nothing, they certainly cannot be considered as an effective detection device in place of smoke alarms in terms of life safety in residential occupancies today. Smoke alarms should be the primary detection device in the sleeping areas and most other areas of the home. Heat detectors can be used effectively in areas where smoke detectors would be subject to false alarm, such as

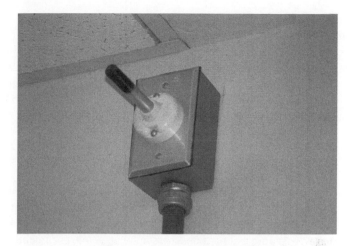

FIGURE 3.26 Heat detectors were the first fire detection devices developed.

kitchens, bathrooms, basements, and garages. Heat detectors are often substituted for smoke detectors in commercial occupancies where a smoke detector would be subject to false alarms. Such locations might include kitchens, mechanical areas, electrical rooms, outdoor elevator lobbies in parking garages, and other locations subject to dust, steam, and ambient conditions that would cause smoke detectors to malfunction. Heat detectors are also used in conjunction with sprinkler heads in elevator operations to activate elevator safety components to prevent water damage to electrical components from sprinkler operation (this will be explained further in chapter 7, "Elevators and Controls").

Two basic types of heat detectors are in common use today, the fixed-temperature and the rate-of-rise. Fixed-temperature heat detectors are intended to go into alarm when the temperature in an area reaches a predetermined level. Temperature ratings start at around 135°F (58°C) and go up from there. Higher temperature heat detectors are used in locations where ambient temperature is normally high, such as steam rooms, boiler rooms, and autoclave rooms. Heat detectors may also be used in open and closed parking garage elevator lobbies, where smoke detectors would be subject to false alarm due to ambient conditions.

Rate-of-rise heat detectors work on the principle that flaming fires will raise the temperature of the air in a room rapidly. Where fixed-temperature heat detectors do not react until the temperature in an area reaches their predetermined fixed temperature, rate-of-rise heat detectors respond to rapid changes in temperature in the room. Typical rate-of-rise values start at a 12 to 15°F (7 to 8°C) rate of rise in air temperature per minute. When placing these detectors, be sure to avoid locations where such fluctuations in temperature may occur naturally, such as in kitchens, laundries, dishwashing areas, autoclave areas, truck repair facilities, or near heating appliances. Rate-of-rise detectors will respond to changes in temperature regardless of where the temperature starts. It could be freezing in an area and just inducing heat into the space could trigger a rate-of-rise response from the heat detector. Detectors also exist that contain both the fixed-temperature and rate-of-rise features. Whichever event occurs first, the detector will respond to either the predetermined temperature or the rate of rise. Certain conditions may result in either element responding first. Rate-of-rise will respond quickly to a rapidly developing fire. Fixed-temperature detectors will respond to a slowly developing fire, such as one with limited O_2 causing a smoldering fire.

FLAME DETECTORS

Other sophisticated types of detection devices are designed for use in special circumstances; however, these detectors are very expensive and are limited to certain types of operations and are beyond the scope of this text. Few jurisdictions will encounter these specialized detectors. Those that do have them will need to address them as needed with training of response personnel and code enforcement.

Regardless of the type of fire alarm system, activation devices, detection devices, manual pull stations, or other devices used, it is important that the fire department review and approve the plans for design and code compliance and that firefighters become familiar with the operations of the components.

AUDIBLE DEVICES

Fire alarm systems may have alarm bells, audible tones, horns, or voice initiation devices installed (figure 3.27). Audible tone devices may involve many different types of sounds, including bells, depending on the manufacturer. However, within any given occupancy, there can only be one type of fire alarm audible signal in public areas. In buildings where people may not all be evacuated from a building when

FIGURE 3.27 Fire alarm systems may have alarm bells, audible tones, horns, or voice initiation devices installed.

a fire alarm sounds, such as hospitals, sometimes coded bell systems are used to alert staff to a potential fire. Certain configurations of bells will indicate the floor, wing, and room areas of fires that will only be known to staff. Staff members will then respond to those areas to render assistance where needed. Procedures often call for staff to bring fire extinguishers with them as well. Sounds produced by audible devices should be not less then 15 dB above ambient sound levels and not more than 120 dB at any point in the occupancy. Part of the final acceptance test of any newly installed fire alarm system should include taking readings of sound levels with a sound meter. Areas such as mechanical rooms and other rooms where there is a lot of ambient noise may require the installation of specialized audio and visual appliances. Audio devices are generally mounted according to manufacturer's instructions. Conditions in a facility may, however, require additional devices to be installed to provide proper sound coverage throughout. It may not be necessary to install all of the devices shown on the fire alarm shop drawings. Given some experience, you can get a pretty good idea how many audio devices will do the job. You can install too many audio devices in a building. One such facility at the University of Maryland had so many devices that the sound was unbearable. One hundred thirty audible devices were removed from the building and the remaining sound level was still adequate to alert occupants, although some people still complain about the noise. Buildings with voice fire alarm systems have devices that are adjustable between ¼ and 1–4 watts. Following installation of the system during testing there may be locations where the voice part of the system is not loud enough. The sound can be increased by increasing the speaker wattage. However, care has to be taken in the design of the fire alarm system to provide an amplifier for the voice system large enough to allow for maximum output on speaker devices. Otherwise, the wattage on the speakers can not be adjusted.

Visual Devices

Visual devices are also required in buildings with fire alarm systems. In older buildings, the visual devices may be nothing more than flashing lights covered with red plastic bezels that have the word "Fire" on them. Modern systems, however, are outfitted with strobe lights. Strobe lights come in varying levels of intensity depending on the area in which they are to be installed (figure 3.28). Installation recommendations are found in NFPA 72 for spacing in corridors and numbers and candela required per area. Strobe lights may be installed on walls or ceilings but there are requirements for the distance from the ceiling if placed on the walls. Usually the requirement is for a maximum of eighty inches above the floor. The bottom line is that there must be a visual device that can be seen from any location in the egress system and in any room greater than four hundred square feet.

Voice Alarm Systems

Voice alarms automatically send a voice evacuation message to speakers in selected areas of high-rises or expansive buildings, hospitals, and other buildings where total evacuation is unwanted or impractical (figure 3.29). They are also used in high-rise and other buildings where total evacuation is preferred. A typical high-rise

FIGURE 3.28 Strobe lights provide a visual notification of fire and come in varying levels of intensity depending on the area in which they are to be installed.

arrangement would provide for the following areas to automatically receive a prerecorded evacuation signal: The floor where the alarm originates and the floors above and below it. Arriving firefighters can evacuate additional areas by manually activating one, multiple, or all floors with the manual select switches in the fire command center. The voice paging capability can also be used without the fire

FIGURE 3.29 Voice alarms automatically send a voice evacuation message to speakers in selected areas of high-rises or expansive buildings, hospitals, and other buildings where total evacuation is impractical.

FIGURE 3.30 Arriving firefighters can evacuate additional areas by manually activate one, multiple, or all floors with the manual select switches in the fire command center. Adjacent to each manual select switch, visual indicators show which evacuation zones are activated at any given time.

alarm being activated (figure 3.30). This can be useful when you have a chemical spill or leak that does not require the evacuation of all the building floors. You can select which areas need evacuation and give a voice message over the system to alert occupants. Firefighters also can override the prerecorded fire alarm message and broadcast live voice announcements to any or all evacuation zones with a microphone at the command center. Paging can be used to give special instructions or information to persons evacuating the building or people in areas of rescue assistance. Adjacent to each manual select switch, visual indicators show which evacuation zones are activated at any given time. Arrangement of evacuation zones depends upon the design of the building and any evacuation plan in place. Each floor is typically one evacuation zone. Areas that are not separated by fire or smoke barriers should not be divided into multiple evacuation zones. However, if a floor is divided by fire or smoke barriers to enable occupants to take refuge on either side of the barrier, multiple evacuation zones should be provided. Operators at the fire command center will only be able to give different verbal instructions to those on either side of the barriers if the zone boundaries coincide with the rated barriers. In addition to normally occupied spaces, most building and fire codes require speakers in stairways and elevator cabs. Each stairway and each bank of elevators should comprise a single evacuation zone. In a building with selective evacuation, it is undesirable to automatically activate the speakers in these areas. Also, there are typically no detectors to warn of fire or smoke within the stairways or elevator cabs. Each of these zones usually has "manual only" selection capability for the operators in the fire command center. If a stairway has detectors, the speakers in that particular stairway could be configured into a separate, automatically activated

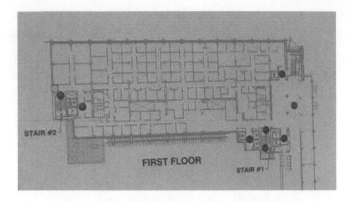

FIGURE 3.31 Floors that are physically open to one another should be arranged as a single evacuation zone.

evacuation zone. Designers should ensure that evacuation signals are not heard in areas that are not to be evacuated, such as in three-floor evacuation, mezzanines, and atriums.

Floors that are physically open to one another should be arranged as a single evacuation zone (figure 3.31). This avoids the confusion possible when occupants in portions of the space hear an evacuation signal but cannot clearly decipher it. A common example of this situation is a series of parking garage levels connected by open ramps. The group of interconnected levels should be designated as a single evacuation zone on the "floor, floor above, and floor below" automatic evacuation scenario. Atriums and other large open spaces spanning multiple floors also deserve special attention in buildings with selected evacuation. The arrangement depends upon the egress requirements and the building's evacuation plan. The entire atrium should comprise one evacuation zone. It may be desirable to activate only the atrium zone upon receipt of an alarm signal from within the atrium and not from alarm signals in other areas. Designers should consider the legibility of signals in areas adjacent to the atrium in order to avoid occupant confusion.

FIRE DEPARTMENT COMMUNICATIONS SYSTEMS

Fire department communications systems are two-way telephone systems typically required in high-rise buildings. The fire command center contains the control unit with the main handset for use by the fire department commanders (figure 3.32). Either handsets or jacks for handsets are then placed in areas of the building for firefighters to communicate with other firefighters in the fire command center (figure 3.33). If the system uses jacks, a number of portable phone handsets with plugs are provided in the fire command center for distribution to firefighters (figure 3.34). Designers should plan for handsets or jacks in locations where firefighters are likely to be operating. NFPA 72 requires only one handset or jack per floor, one per exit stairway, each elevator car, and each elevator lobby. The IBC currently requires handsets or jacks in the same locations as NFPA 101 and also in standby power rooms,

FIGURE 3.32 The command center contains the control unit for the firefighter phone system with the main handset for use by the fire department commanders.

fire pump rooms, and areas of refuge. These additional jacks or handsets can provide more rapid communications from these critical areas. Both the IBC and NFPA 101 contain exceptions that allow fire departments to approve their radio systems as a substitute for two-way firefighter telephone systems in a high-rise building. For a radio system to be equivalent, the radio signals should be operable in the same areas

FIGURE 3.33 Either handsets or jacks for handsets are then placed in areas of the building for firefighters to communicate with the command center.

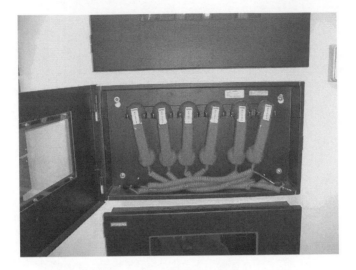

FIGURE 3.34 If the system uses jacks, a number of portable handsets with plugs are provided in the command center for distribution to firefighters.

(the fire command center and each remote jack or handset location). To exercise this option, designers or building owners should test radio signals and provide documentation of successful results. Signal retransmission devices may be necessary to provide optimum radio communications inside of some buildings. Even if radio communications are good within a building, it may be wise to install a firefighter two-way phone system anyway. Phone systems can be used by handicapped persons in areas of rescue assistance. They can be used during building emergency operations when radio frequencies are tied up with other operations. They can be used for secure transmissions within the building. Additionally, when bomb threats or bombs are thought to be present in a building, firefighter phones can allow for continued communications when radios cannot be used because of their potential to set off explosive devices controlled by radio.

FIREFIGHTER RETRANSMISSION SYSTEMS

Fire department portable radios are frequently unreliable inside buildings and other structures such as tunnels. Construction materials, earth, and changes in the radio-frequency environment can greatly reduce the strength of radio signals. If a firefighter inside is unable to transmit or receive, he or she must relocate closer to an exterior opening, move to a different floor, use an alternate means of communication, or resort to runners or direct voice communications. Cell phone signals are affected by the same factors as radio signals. Landline phones will allow firefighters to communicate with dispatchers but not other units; they may also be affected by the incident occurring in the building. All of these factors may delay operations and create greater challenges in maintaining crew integrity. New technology can improve signal transmission within buildings and structures through fixed communications infrastructures.

Passive approaches simply provide a conduit to assist in the transmission of signals. However, active methods involve powered devices to amplify and retransmit signal. For example, the "passive antenna system" includes both an internal and an external antenna, connected with a short coaxial cable. A "radiating cable," also known as a "leaky coax," is a network of coaxial cables with slots in the outer conductor that create a continuous antenna effect.

Increasing in popularity is an active signal transmission method involving a signal booster also known as a "bidirectional amplifier," or simply BDA. These powered devices amplify signals between an external antenna and one or more internal antennae. Both reception and transmission are amplified messages on portable radios within the building. A network of antennae placed at strategic locations or a leaky coaxial cable distributes signals throughout the coverage area. Some installations combine passive and active approaches. Passive antennae generally work well in small, well-defined areas. BDAs function well in large diverse areas that need a solution for inadequate coverage. Some jurisdictions have adopted laws or ordinances concerning public safety radio communications. In others, designers should consider specifying a study to determine the possible need for retransmission devices. Installing stationary communications infrastructures in high-rise buildings and tunnels is one way to resolve communication problems like those encountered by the fire department of New York on September 11, 2001. Without laws requiring this equipment, cost considerations may discourage owners from voluntary installations. However, owners of high-rises or other target hazards may be swayed by increases in property value or improved safety for tenants. Alternatively, in high-rise buildings where firefighter communication systems are required, a code official may permit the substitution of fixed communications infrastructures. Perhaps, in the future, insurance companies will offer lower premiums for buildings containing such installations. Many local communications ordinances currently in effect in the United States contain specific requirements for system performance. These include signal strength, area coverage, reliability, secondary power supply, interference filters, acceptance testing upon completion, and ongoing periodic testing.

FIRE COMMAND CENTERS

Building or fire codes typically require high-rise buildings to have a dedicated room or other location containing the fire alarm and related fire protection control equipment (figure 3.35). These are called "fire command centers (FCC)" in NFPA 72, and the IBC. The term "emergency command center" is used in the 2006 version of NFPA 101, the *Life Safety Code*, and NFPA 1. Industry also uses the expression "fire control room." Both the IBC and NFPA 72 require the room containing the fire command center to be one-hour fire rated. These rooms often have exterior entrances, which should be prominently marked. The IBC requires the room to be at least ninety-six square feet, with a minimum dimension of eight-feet. NFPA 72 requires at least a three-foot clearance in front of all control equipment. The IBC contains a comprehensive list of equipment required in a fire command center. The lists in NFPA 1, NFPA 101, and NFPA 5000 are about half as long, and all these items are

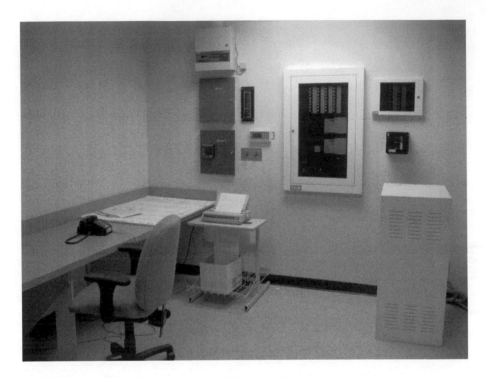

FIGURE 3.35 Building or fire codes typically require high-rise buildings to have a dedicated room or other location containing the fire alarm and other related fire protection control equipment.

on the list in the IBC as well. The following equipment should be located in the fire command center:

- Emergency voice/alarm communication system unit
- Fire department communications unit
- Fire detection and alarm system annunciator panel
- Annunciator unit visually indicating the location of the elevators and whether they are operational
- Status indicators and controls for air-handling systems
- The firefighter's control panel for smoke control systems
- Controls for unlocking stairway doors simultaneously
- Sprinkler valve and water flow detector display panels
- Emergency and standby power status indicators
- Telephone for fire department use with controlled access to the public telephone system
- Fire pump status indicators
- Schematic building plans
- Worktable
- Generator supervision devices, manual start and transfer features
- Public address system

FALSE ALARM REDUCTION

One of the leading causes of firefighter death and injury is responding to and from alarms. One of the best ways to reduce those deaths and injuries after appropriate safety measures is reducing the number of alarms responded to. While this may be unpopular to those wanting to establish job security, we have enough to do, and reducing false alarms will only serve to free us up to do those other things. Many departments send a reduced response to building fire alarms anyway. False alarms can be prevented! I know that sounds like an empty campaign slogan; however, it is a fact. It requires effort to determine what is causing the false alarms and then a commitment to make changes. Manual pull stations are frequently a cause of false alarms. People bump into them, younger kids are attracted to them, older kids pull them for pranks—there are numerous reasons they are activated besides the reason they are designed for. These issues can be stopped or greatly reduced by placing protective covers with alarms over the pull stations. Oftentimes false alarms occur because some fire alarm system component or components are improperly placed within the building. Fire alarm shop drawings should be carefully examined to determine proper device placement. Smoke detectors are one of the leading sources of false alarms in buildings other than residential occupancy. Smoke detectors were designed for places where people sleep. They are not intended to substitute for warning devices in mechanical and other similar locations to warn when equipment malfunctions or fails. There is a section of NFPA 72 that discourages this type of use. Smoke detectors should not be placed in locations where dirt, weather, or ambient conditions would subject the smoke detectors to false alarm. Smoke detectors in some jurisdictions are used as an equivalency to sprinkler system requirements. I am not saying that I agree, but it happens. In smaller communities and rural areas, when hospitals, nursing homes, and schools should be sprinkled, smoke detectors are often allowed as an equivalency. When this occurs, just be careful with the placement of devices so that they do not cause false alarms. Do not put them in or near bathrooms or areas where food is prepared. Toasters and microwaves are notorious for producing products of combustion that cause smoke detectors to go into alarm. Heat detectors would be much better alternatives in these types of locations. Heat detectors can also be the source of false alarms if they are not properly placed, the right type, or at proper temperature settings for the surroundings. Steam has a temperature of 212°F at normal pressures. It makes no sense to place a normal temperature heat or smoke detector in an area where there is a source of steam, which is released into the atmosphere. Yet it occurs frequently. Beam detectors are used in large open areas where detection is desired but smoke detectors would not function well. These detectors have a device that emits a beam of light across the open space to a receiver. When smoke breaks the beam of light, the device and, subsequently, the fire alarm system go into alarm. Unfortunately, anything else that breaks the beam of light will also send the system into alarm.

Sprinkler system devices that are hooked into the fire alarm system such as flow switches can cause false alarms. Flow switches have adjustments on them. If they are adjusted to be too sensitive, normal fluctuations in water pressure can cause them to go into alarm. I would recommend that flow switches be adjusted to the least sensitive setting. This will not change the response time for sprinklers that much when a fire occurs, but it will reduce the number of false alarms. This will delay slightly the time it takes to activate the building fire alarm system, but the sprinkler system will already be controlling or extinguishing the fire, so I believe it is a good trade-off. It is important to remember that it does no good to have a fire alarm system if it constantly results in false alarms and the building occupants start to ignore the system when it is activated.

We had a serious false fire alarm problem at the University of Maryland, Baltimore. In some buildings it became so serious that people ignored fire alarms when they were activated because they thought it was just another false alarm. In an attempt to reduce false alarms on campus, we conducted an analysis of the causes of false alarms. Once the major causes were determined, we prioritized those causes we thought we could correct. By far the most significant cause was contractors working in campus buildings. Training classes were developed and contractors were required to attend. During the classes, contractors were educated about the common false alarm causes and how to avoid setting off an alarm while working on campus. The university's Hot Work Program was strengthened and enforcement was emphasized. When contractors obtained hot work permits they were "lectured" about preventing fires and false alarms. When a false alarm was caused by contractors they were lectured again. Efforts to reduce contractor-caused false alarms were successful. Contractor-caused false alarms improved from the number one cause to zero contractor-caused false alarms in just a year. Education and training for false alarm prevention was extended to university maintenance personnel as well with equally successful results.

The next focal point of false alarm reduction was the fire alarm systems themselves. NFPA 72, the *National Fire Alarm Code*, allows certain building fire alarm devices to be placed on supervisory signal. Many of these devices were major causes of false alarms in university buildings. Wherever possible, we placed duct smoke detectors on supervisory signal. We placed elevator machine room smoke and heat

TABLE 3.1

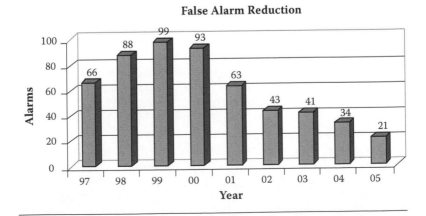

detectors on supervisory signal. Beam detectors were placed on supervisory signal. These efforts were successful in reducing false alarms. Improper use and placement of smoke detectors was also a major cause of false alarms. Smoke detectors in basement corridors of buildings, in building mechanical areas, and on loading docks were removed or changed to heat detectors. Smoke detectors in elevator lobbies in parking garages and on loading docks were changed to heat detectors. Sprinkler flow switches were set to the least sensitive settings. The results were dramatic in terms of reducing false alarms while not impacting the life safety of building occupants because most of our buildings are fully sprinkled. Placing alarmed covers over manual fire alarm pull stations on the first floor of buildings and in problem areas where false alarms have occurred in the past has also been very successful. Several fire alarm system replacements and upgrades were accomplished, which helped with the reduction of false alarms as well.

Following a five-year concerted effort to reduce false alarms on campus we have experienced a 79 percent reduction in the total annual false alarms between 1999 and 2005 (table 3.1). This dramatic reduction has also resulted in better participation in drills when they are conducted and better responses to actual emergencies when they have occurred.

4 Suppression Systems

INTRODUCTION TO SUPPRESSION SYSTEMS

Preventing fires from happening is the preferred course of action to take toward fighting fire. Fires can be prevented through enforcement of building and fire codes, public education, and inspection programs. If, however, fire does occur despite our best efforts to prevent it, it is important to have a fire suppression system in place to contain or extinguish the fire before the fire department arrives. Sprinkler systems and other suppression systems such as clean agents, halon, dry chemical, and carbon dioxide provide early fire control or extinguishment, helping to mitigate the hazards for occupants and firefighters alike. Building codes, fire codes, and life safety codes specify when to provide sprinkler or other type of suppression system. Sprinkler requirements may be located in either locally written codes or adopted model codes such as the IBC, the IFC, NFPA 1, NFPA 101, or NFPA 5000. In addition, various sections of the Occupational Safety and Health Administration (OSHA) standards require the installation of sprinkler systems in certain types of occupancies. A widely accepted installation standard for commercial system design is NFPA 13, *Standard for the Installation of Sprinkler Systems.* Other standards include: NFPA 13D, *Standard for the Installation of Sprinkler Systems in One- and Two-Family Dwellings and Mobile Homes*; and NFPA 13R, *Standard for the Installation of Sprinkler Systems in Residential Occupancies up to and Including Four Stories in Height.* Designers may also refer to NFPA 13E, *Recommended Practice for Fire Department Operations in Properties Protected by Sprinkler and Standpipe Systems*, although any given fire service organization may follow different standard operating procedures.

NFPA 13E provides information on what firefighters should know when responding to facilities that have interior sprinkler systems, exterior sprinkler systems, and standpipe systems. Training programs should be developed on the fundamentals of automatic sprinkler and standpipe systems using this standard to prepare firefighters to deal with these systems in buildings. Standard operating procedures (SOPs) should be developed as well, and this standard can be used as a resource for those procedures. SOPs should be specifically developed for fireground operations in sprinkled buildings. Sprinkler systems in buildings do not always function as intended, generally because of human error or lack of code enforcement and maintenance of systems. Firefighters should be able to deal with situations where there is unsatisfactory sprinkler performance. These include a closed valve in the water supply line, inadequate water supply to the sprinkler system, and occupancy changes that have rendered the installed sprinkler system unsuitable. Sprinklers have to be unobstructed in order to function properly as well. NFPA 13 requires an eighteen-inch clearance under sidewall, extended coverage upright and pendent sprinkler heads so that the discharge pattern will not be obstructed. Early suppression fast response (ESFR) sprinklers require a thirty-six-inch clearance below the deflector. Standard pendent or upright spray sprinklers do not require the eighteen inches for storage along a wall. Fire prevention personnel and responding firefighters should notify building owners when sprinklers are found to be obstructed and have the obstructions removed.

Alternative suppression system standards include NFPA 11, *Standard for Low, Medium, and High Expansion Foam*; NFPA 11A, *Standard for Medium and High Expansion Foam Systems*; NFPA 12, *Standard for Carbon Dioxide Extinguishing Systems*; NFPA 12A, *Standard on Halon 1301 Fire Extinguishing Systems*; NFPA 15, *Standard for Water Spray Fixed Systems for Fire Protection*; NFPA 16, *Standard for Installation of Foam-Water Sprinkler and Foam-Water Spray Systems*; NFPA 17, *Standard for Dry Chemical Extinguishing Systems*; NFPA 17A, *Standard for Wet Chemical Extinguishing Systems*; NFPA 96, *Standard for Ventilation Control and Fire Protection of Commercial Cooking Operations*; and NFPA 2001, *Standard on Clean Agent Fire Extinguishing Systems*. There is some flexibility written into portions of extinguishing system standards that may impact the fire service. It is important for sprinkler designers and fire alarm designers to work together, especially in unusual buildings. The fire alarm system will often have an annunciator to indicate the location of the alarm to the fire department. Sprinkler piping arrangement will limit options for the fire alarm annunciation of water flow signals. Coordination is essential to furnish the fire service with clear information on the fire or its location.

Sprinkler designers often think in terms of ceiling levels, since sprinkler piping and sprinkler heads usually are at ceilings or roof decks. However, alarm signals are reported in terms of their floor level to enable the fire department to respond to the correct floor during an emergency. Consider the situation of a building with two levels adjacent to a single-level "high-bay" area. The first-floor sprinkler zone should include both the high-bay area and the lower level of the two-level section because each of these areas shares the same floor. In buildings with standpipe systems, sprinkler systems are usually combined with them and fed from the bulk feed mains or from vertical standpipe risers (figure 4.1). NFPA 13 requires that sprinkler controls remain independent of standpipe systems. Typically, all sprinklers would be located downstream from a control valve that will not shut off any fire hose connections (figure 4.2). This enables the fire department to shut off the sprinklers during the rare occasions when a sprinkler pipe fails or the sprinklers are not controlling the fire. They can also shut off the water when the fire is extinguished to reduce water damage. In this manner, hose connections remain available for manual fire suppression without losing pressure from the broken pipe or the excessive number of activated sprinklers. In some situations, when a building does not include a standpipe system, NFPA 13 allows fire hose connections to be fed from sprinkler systems provided that there is enough water supply. In these cases, closing the sprinkler system valve would shut off the fire hose connections. In some cases, sprinkler systems are fed from two different standpipes or feed mains, in a dual-feed arrangement. Although this provides a hydraulic design advantage, NFPA 13 recommends against it to avoid confusion. If a designer chooses this arrangement (and the code official permits it),

FIGURE 4.1 In buildings with standpipe systems, sprinkler systems are usually combined with them and fed from the bulk feed mains or from vertical standpipe risers.

cross-reference signs should be provided at each valve. Each of these signs would indicate the location of the companion valve that feeds the same system. No single sprinkler system should be fed from three or more points, since the flow from a single sprinkler may not activate any of the flow switches.

Fire service personnel often need rapid access to water supply valves. If a valve is closed during an incident, it may need to be opened to permit flow of water.

FIGURE 4.2 Typically, all sprinklers would be located downstream from a control valve that will not shut off any fire hose connections.

FIGURE 4.3 NFPA 13 requires marking for all water supply control valves, including main valves, pump valves, sectional valves, and zone valves.

If a sprinkler valve is open, it may need to be closed to assist in manual suppression efforts. NFPA 13 requires marking for all water supply control valves, including main valves, pump valves, sectional valves, and zone valves (figure 4.3). The wording "control valve" by itself does not tell a user the specific use of the valve or what portion of the system is downstream of a particular valve. Firefighters can tell by looking at the valve that it is a control valve and a sign is not needed to tell them. Using more descriptive labels such as "12th Floor" or "Pump Bypass" will avoid confusion and provide firefighters with needed information. If valve identification is not obvious, an additional diagram should be provided. For instance, if a floor has multiple zones, each control valve sign should identify the corresponding zone, such as "12th Floor East" or "Zone 7-2." A diagram of zones and the boundaries between them should be mounted adjacent to each valve. This will enable firefighters to quickly determine which valve controls each specific area. NFPA 13 requires valves to be accessible for operation. If valves are located in stairs, they will be protected and easily accessible during a fire event. When a water supply control valve must be located in a room or in a concealed space, a sign outside the door or access panel helps firefighters to quickly locate it (figure 4.4). If the concealed space is above a suspended ceiling, the appropriate place for the sign is on the fixed ceiling grid, rather than on a removable ceiling tile (figure 4.5). In addition, some jurisdictions require exterior signs that indicate the locations of interior valves. Valve handles are often located high enough to be out of vandals' easy reach. However, such placement requires a ladder to reach them when necessary. Although some jurisdictions may require that valves be low enough to reach without a ladder, all minimum height requirements for obstructions must be followed. Valves for testing and draining purposes should also be labeled. This will prevent any potential confusion.

FIGURE 4.4 When a water supply control valve must be located in a room or in a concealed space, a sign outside the door or access panel helps firefighters to quickly locate it.

Exterior valves should be placed in locations accessible even during a fire incident. Wall-mounted valves should be positioned no higher than five feet above grade (ground level) and located at least forty feet from openings such as windows, doors, or vents. Post indicator valves should be at least forty feet from the buildings they serve (figure 4.6). The forty-foot distance is called for in NFPA 24. Designers should require proper notification when their designs require systems, or portions of systems, to be temporarily shut off. This would typically occur during system alterations or phased

FIGURE 4.5 If the concealed space is above a suspended ceiling, the appropriate place for the sign is on the fixed ceiling grid, rather than on a removable ceiling tile.

FIGURE 4.6 Exterior post indicator valves should be at least forty feet from the buildings they serve.

installations. In these instances, the design documents should require notification of any system impairments to the responsible fire service organization and coordination with the fire service about any requirements that these impairments may entail.

NFPA 13 requires installation of sprinklers throughout the building. However, in some situations the code or standard requiring sprinklers calls for protecting only a portion of a building. In these cases, exterior signage should indicate the portion of the building covered (figure 4.7). A good location for this sign would be at the fire department connection. Residential sprinkler systems installed under NFPA 13D and 13R primarily protect lives rather than property. Since property protection is secondary, large and significant areas may not have sprinkler protection. One- and two-family houses protected by NFPA 13D systems are readily recognized as having this partial, life safety type of protection. Apartments and condominiums with NFPA 13R systems

FIGURE 4.7 In cases where a sprinkler system only covers a portion of a building, exterior signage should indicate the portion of the building covered.

may not be easy to identify. These systems are allowed in buildings four stories or less in height. However, some buildings that are considered four stories in height by building codes may still contain additional levels, such as lofts, and basements that may be partially below grade. Several sides of these buildings may have six occupied levels above grade and still be considered four stories in height. The large unsprinkled areas can adversely impact firefighter safety and consequently the tactics employed. Fire department ground ladders may not reach the top occupied stories, and some apartment units may not be reached by the available access for aerial ladders. Exterior signage near the fire department connection can alert the fire department to this.

Recent fires, including The Station nightclub fire in Warwick, Rhode Island, on February 20, 2003, that killed one hundred people, further underscore the importance of having suppression systems in place in occupied buildings. Sprinkler systems, the most common type of suppression system, have been around for over one hundred years. Henry Parmalee, an American, invented the fire sprinkler system in 1874 to protect his piano factory. In the beginning, fire sprinklers were generally used to protect property, especially factories and warehouses. Following large loss-of-life fires at the Cocoanut Grove Nightclub in Boston in 1942 where 492 died (see Boston Globe news archives at www.boston.com); the LaSalle Hotel in Chicago in 1946 where 61 people died; and the Winecoff Hotel in Atlanta, Georgia, on December 7, 1946, where 119 people lost their lives, additional sprinkler requirements were placed in the codes. Investigations showed that sprinkler systems in factories and warehouses had a remarkable record in terms of life safety compared to buildings that were not sprinkled. Unfortunately, they have not gone far enough with sprinkler system requirements in all types of occupancies.

The new 2006 version of NFPA 101, the *Life Safety Code*, has been updated as a result of several fires that have occurred, including the Rhode Island nightclub fire.

Additional requirements were added to the assembly occupancy section. Where bars with live entertainments, dance halls, discotheques, nightclubs, and assembly occupancies with festival seating have more than one hundred occupants, they must be protected throughout by a complete automatic sprinkler system. It is curious to note that The Station nightclub fire in Rhode Island had a loss of life of one hundred occupants; however, there were several hundred people in the building at the time of the fire. This version of the *Life Safety Code* also makes sprinkler installation mandatory in single-family homes for the first time. Unfortunately, many jurisdictions have amended the requirement out of the code when they adopted it.

FIRE SPRINKLER MYTHS

Myth: Sprinklers cause water damage.

Fact: Sprinkler systems control a fire in its early stages and limit the growth of the fire. Water damage from the sprinkler system will be much less than the smoke and fire damage if the fire were allowed to grow uncontrolled. This uncontrolled fire growth would, in addition to the smoke and fire damage, have significant water damage from firefighters hoses used to extinguish the fire. Quick response sprinkler heads release thirteen to twenty-four gallons of water per minute compared to over 125 gallons per minute average per firefighter hose line used. Uncontrolled fires significantly increase the damage versus fires where a sprinkler system is in place and controlling or extinguishing the fire.

Myth: If one sprinkler head activates, they will all activate.

Fact: Only one sprinkler head over the point of fire origin will activate initially. Generally, sprinkler heads activate when the temperature at a given head reaches 165°F or whatever the temperature rating on the sprinkler head. Other sprinklers in the area do not activate until their temperature reaches the same level. When one head activates, it usually controls the fire and reduces the room temperature so that other heads do not activate. Only heads in the room where the fire is located will activate under normal circumstances.

Myth: Sprinklers may go off by accident.

Fact: If you are a betting person, the odds are 1 in 16 million per year that the sprinkler system has been installed against an accidental activation. Sprinkler systems are tested and maintained periodically to further reduce the chance of accidental discharge. You have a better chance of being struck by lighting than your sprinkler system going off accidentally!

Myth: Sprinkler systems are ugly.

Fact: Advances in technology have improved the appearance of sprinkler heads. However, what is more important—how they look or how they save lives?

Myth: What about the sprinkler system freezing?

Fact: Sprinkler piping is usually installed on interior walls, eliminating their exposure to freezing temperatures. In cold climates, piping should be insulated against freezing.

Myth: A smoke detector is enough protection.

Fact: Smoke detectors save lives by providing an early warning system for smoke from a fire, but they do nothing to extinguish or control the growth of a fire. Oftentimes smoke detectors are not tested and batteries not replaced or have been removed. Detectors must be maintained in good working order if they are expected to help save your life during a fire.

Myth: I can't afford a sprinkler system.

Fact: In terms of protecting your family from the dangers of fire, you cannot afford not to have a sprinkler system. Installation of sprinklers averages about one dollar per square foot of building size. Many times this represents an increase in construction cost of only one percent of the project cost. Not much to pay for such a high level of protection!

Suppression systems may be wet or dry and composed of extinguishing agents, including water, chemical, gas, liquid, powder, or a combination. The most important thing to determine is which system will be most effective against the hazard being protected against. Suppression systems are designed to extinguish or control a fire until the fire department arrives. Sprinkler systems save lives. They have been around for over one hundred years and have an impeccable record for saving lives and property. NFPA reports that there has never been a multiple loss-of-life fire in a fully sprinkled building. Note: NFPA considers multiple loss of life to be greater than three persons.

SPRINKLER SYSTEMS

Sprinkler systems are designed to respond at predetermined temperatures, automatically discharging a stream of water and distributing it in designed patterns and quantities over selected areas when a fire occurs. A sprinkler system is like having a series of heat detectors spread throughout an area that not only detect heat but react to the heat by fusing and setting off the fire alarm while extinguishing or controlling a fire. It is also like having a firefighter standing by ready to fight a fire when it first breaks out, the point when it is most easily extinguished. There are four general types of automatic sprinkler systems available, wet pipe, dry pipe, pre-action, and deluge. Selection of the proper type for the application and hazard is extremely important. For example, because wet systems contain water, they should not be used in areas where they would be exposed to freezing temperatures. Dry systems, on the other hand, respond slower than wet systems and should only be used where freezing temperatures are encountered. Sprinkler system installation requirements are found in the local building code, the NFPA *Life Safety Code*, and local codes and ordinances. Sprinkler systems are designed using two general types of design criteria, pipe schedule method and hydraulic design method using a computer program. The hydraulic design method is by far the most common method used today. Details on both methods can be found in NFPA 13, *Standard for the Installation of Sprinkler Systems.*

Wet pipe sprinkler systems are by far the most common type of suppression system available (figure 4.8). By code, buildings can be outfitted with full or partial

FIGURE 4.8 Wet pipe sprinkler systems are by far the most common type of suppression system found in sprinkled buildings.

sprinkler systems, although it makes no sense to me to partially sprinkler a building. Buildings should be 100 percent sprinkled. Some of the components necessary for a sprinkler system to be installed and function correctly include the alarm check valve, main drain, main valves, floor zone valves, inspectors test valve, OS&Y valves, water flow switches, and valve tamper switches. Which components are required will depend on the size of the system. Systems can be as simple as one room to complex high-rise installations. In order for a sprinkler system to work effectively, it must have a dependable, adequate water supply. Water supplies can be public water mains, private water mains, water tanks, water towers, or reservoirs. There must also be adequate pressure to operate the sprinkler system. If adequate pressure is not available, then a fire pump will need to be installed (figure 4.9). Tests will need to be conducted at the city water main (fire hydrant) to determine the available water for new sprinkler system installations. Hydraulic calculations for sprinkler systems are generated from computer programs from input of data including the flow tests from the water source.

Dry pipe sprinkler systems are designed for locations where piping will be exposed to freezing temperatures, which can burst the pipe (figure 4.10). Examples of such locations would be high-rise parking garages, underground parking garages, unheated attics, loading docks, unheated rooms or buildings, and all other exterior installations where no heat is present. Antifreeze solutions can also be used in piping subject to freezing and must be installed in accordance with the guidance provided in NFPA 13. Unlike wet sprinkler systems, dry systems do not have water in vulnerable, exposed sprinkler piping. A dry pipe valve located in a heated area

FIGURE 4.9 If adequate pressure is not available, then a fire pump will need to be installed to supply pressure to the sprinkler and standpipe systems.

keeps the water from entering the piping by the use of air pressure (figure 4.11). Air pressure has to be greater than the water pressure being exerted on the dry valve in order to maintain the valve in a normally closed position. Manufacturer's instructions should be followed when pressurizing the dry pipe valve. In the absence of

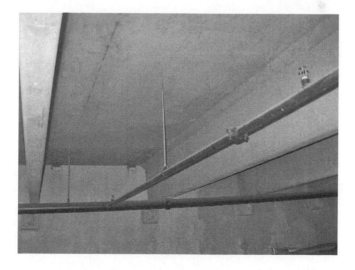

FIGURE 4.10 Dry pipe sprinkler systems are designed for locations such as parking garages and loading docks where piping will be exposed to freezing temperatures, which can burst the pipe.

FIGURE 4.11 A dry pipe valve located in a heated area keeps the water from entering the piping by the use of air pressure.

manufacturer's information, the air pressure should be maintained at a minimum of 20 psi above the trip pressure. Air pressure can be provided by an on-site air compressor or using bottled air (usually nitrogen) to charge the system. Whatever system is used to supply air, it must be able to restore normal air pressure in the system within thirty minutes. In refrigerated spaces below 5°F (15°C), restoration of air pressure must occur within sixty minutes.

Dry pipe sprinkler heads, in general, operate like their wet system counterparts. When a fire occurs and grows to the point that it is producing temperatures that will activate a sprinkler head, the head that reaches its rated temperature first activates. Once this happens, the air is bled off of the sprinkler piping through the open head back to the dry pipe valve. When the pressure is bled off to a point where the air can no longer hold the valve clapper shut, the valve activates and water flows through the piping, forcing the remaining air out of the piping before it is discharged on the fire through the open head. Another option to the dry pipe system in freezing atmospheres is the use of antifreeze solutions in the piping. NFPA 13 gives detailed instructions on the use of that type of system.

Pre-action fire sprinkler systems resemble a dry pipe sprinkler system. Piping is charged with air rather than water, just as in the dry pipe system. Air pressure used in conjunction with pre-action systems is generally less than that for dry pipe systems. Water is held back in a pre-action system by a pre-action valve. Activation of the pre-action valve is tied to a supplemental detection system. When the detection system activates because of heat or smoke, the pre-action valve automatically opens and admits water to the sprinkler piping. The sprinkler system now becomes a wet pipe sprinkler system and discharges water from individual heads when those heads reach their rated temperature and activate. Air is kept in the dry pipe portion of the pre-action system to detect leaks in the piping and fittings. If the air leaks out, it sets off a low-pressure alarm. However, no water is released because the air leak does not open the pre-action valve. Pre-action systems are often used in areas where computers and communications equipment are located.

Deluge sprinkler systems are designed to deliver large quantities of water over an extended specified area quickly. Sprinkler heads do not have thermally sensitive elements and are always open. Just as with the pre-action system, activation of the deluge system depends on a supplemental detection system that opens the deluge valve when smoke or heat are detected. The primary difference between the deluge system and pre-action system is the fact that water does not discharge with a pre-action system until a sprinkler head activates from heat. With the deluge system, the system discharges water when the detection system operates. This is a distinct disadvantage if there is a concern for water damage. Deluge systems are generally used in locations where expensive equipment is to be protected and it is not sensitive to water damage. Aircraft hangers are a typical location for deluge sprinkler systems.

OCCUPANCY CLASSIFICATIONS

The sprinkler design, installation, and water supply requirements are determined by the occupancy classification of the building. There are five general occupancy classifications outlined in NFPA 13, Light Hazard, Ordinary Hazard (Group 1), Ordinary Hazard (Group 2), Extra Hazard (Group 1), and Extra Hazard (Group 2); see table 4.1.

TABLE 4.1

Occupancy Classification for Sprinkler Design

Classification	Combustibility	Quantity Combustibles	Stockpiles of Combustibles	Rates of Heat Release
Light	Low	Low	None	Low
Ordinary (Group 1)	Low	Moderate	< 8 ft	Moderate
Ordinary (Group 2)	Moderate to high	Moderate to high	< 12 ft	Moderate to high
Extra hazard (Group 1)	Very high; limited or no flammable liquids	Very high	N/A	High
Extra hazard (Group 2)	Moderate to substantial flammable liquids	Extensive shielding of combustibles	N/A	High

Note: Reprinted with permission from NFPA 101®, *Life Safety Code*®, Copyright© 2006, National Fire Protection Association, Quincy, MA 02169. This reprinted material is not the complete and official position of the National Fire Protection Association on the referenced subject, which is represented only by the standard in its entirety.

TABLE 4.2

Commodity Classifications

Classification	Combustibility	Construction of Containers	Palleted or Not	Examples
Class I	Noncombustible	Wooden pallets; shrink-wrapped; corrugated cartons	With or without	Batteries; fertilizers, bagged; milk; bottles, jars
Class II	Non-combustible	Wooden crates; solid wood boxes; corrugated cartons	With or without	Light fixtures; photographic film; wood products, e.g., plywood, syrup barreled, wood
Class III	5% by weight or volume Group A or Group B plastics	Plastic	N/A	Charcoal, bagged; cereals, packaged, cartoned; furniture, wood, pvc
Class IV	Combustible	Partially or totally of Group B plastics; free flowing Class A plastics	With or without	

COMMODITY CLASSIFICATION

Commodities are classified according to their fire hazards. Most are in bulk form and with or without pallets. Pallet construction may be wood, paper, natural fibers, or plastic. Cartons and containers may be corrugated cardboard or wood. Cartons may be divided or undivided. Some pallets may be shrink-wrapped. There are four commodity classifications; see table 4.2.

CLASSIFICATION OF PLASTICS, ELASTOMERS, AND RUBBER

Plastics, elastomers, and rubber materials are classified as Group A, Group B, or Group C. Requirements for the type, quantity, and storage configuration of combustibles for any commodity classification is basically an effort to identify the likely fire severity, based on its burning characteristics, so the fire can be effectively controlled by the given sprinkler protection for the commodity class; see table 4.3.

TYPES OF SPRINKLER HEADS

Sprinkler heads used in fire sprinkler systems must be listed by a recognized testing laboratory such as UL or FM. When being installed or replaced, only new heads are allowed to be used. Sprinkler systems and sprinkler heads have been protecting life and property effectively since the mid 1800s. Early sprinkler heads, sometimes referred to as conventional sprinklers, discharged water against the ceiling of a room.

TABLE 4.3

Classification of Plastics

Plastics Group Classification	Examples
Group A	Candles Carpet tiles Furniture (with foam plastic cushioning) Plastic containers Mattresses (foam)
Group B	Cellulosics Chloroprene rubber Natural rubber Nylon Silicone rubber
Group C	Fluoroplastics Phenolic PVCPVDC Urea formaldehyde

This type of discharge made it difficult for fire to burn above sprinkler heads. However, because of this type of pattern, they require closer spacing than modern versions of sprinkler heads. During the early 1950s, a new type of sprinkler head was developed with a more efficient umbrella-shaped spray pattern. The primary change resulting in the new pattern was the redesign of the deflector.

Normally, the water in a wet pipe sprinkler system is kept in place in the piping by a cap, plug, or valve held firmly against the discharge orifice by a system of levers and links or other releasing devices applying pressure down on the cap and held in place firmly by struts on the sprinkler. Sprinkler heads release the caps, plugs, or valves when the temperature from a fire in a room reaches a predetermined level. Several systems are used to detect this heat level and respond by releasing the caps. One such system is the use of fusible metal alloys of predetermined melting point. When the temperature reaches the melting point of the alloy, the sprinkler head activates. Another quite common type of release system is the use of a glass bulb filled with liquid. Bulbs are not completely filled with liquid, leaving a small air bubble trapped within. When a fire occurs, the liquid expands from the heat and the bubble is compressed and ultimately absorbed by the liquid. Following the absorption of the bubble the pressure in the bulb rises dramatically, causing the bulb to shatter, thus releasing the cap holding back the water. The liquid in the bulbs is color-coded to reflect the activation temperature of the sprinkler head. Temperature control is accomplished by adjusting the amount of liquid and the size of the bulb when it is sealed. Other methods used, but not quite as common, include the use of bimetallic discs, fusible alloy pellets, and chemical pellets.

The most common types of sprinkler heads are the upright and pendant heads. They are basically the same type of head with different orientations and modified deflectors. Pendant heads are mounted in an orientation that makes them appear to be upside down. Upright sprinkler heads are orientated in the right-side-up position. Sprinkler heads are selected for the particular application present in an area where the sprinkler system is located. Areas with ceilings usually are outfitted with pendant sprinkler heads. Those areas without ceilings are usually equipped with upright sprinkler heads. There are also two basic types of sprinkler heads, those designed to control fires and those designed to extinguish fires (table 4.3). Until recently, automatic fire sprinklers were only designed to control a fire and activate the fire alarm system to notify the fire department so they could respond and finish extinguishment with hose lines. Building owners, designers, and fire code officials need to meet to determine the sprinkler system performance objectives of whether to control or extinguish the fire.

Deflectors on sprinkler heads, in addition to the obvious function, have considerable information about the head listed on the deflector. For one, the recognized testing laboratory symbols are stamped on the head, in most cases UL but also UL Canada and Factory Mutual (FM). The date the sprinkler head was manufactured, sprinkler model number, PEND (indicates a pendant sprinkler), and the operating temperature of the head are also located on the deflector. By looking for this information, inspectors can make sure the proper heads for the application intended are installed.

EARLY SUPPRESSION FAST RESPONSE (ESFR) SPRINKLERS

While sprinkler heads and systems have been around for a long time, several new types of heads have been developed over the past couple of decades. One such development is the early suppression fast response (ESFR) sprinklers. The theory behind fast response sprinklers is that they can be effective in suppression of a fire if the fast response is accompanied by a discharge density capable of reaching through the fire to get at the source of the burning fuel. If suppression of the fire is achieved early in the fire development, it is expected that only sprinklers in the immediate fire area will operate. This sprinkler was first developed to deal with fires in high-rack storage exposures. ESFR sprinklers are generally only used for wet sprinkler systems unless specifically listed for use in dry systems.

SIDEWALL SPRINKLERS

Sidewall sprinklers have the same elements as standard sprinklers but have a different deflector. Water discharges mostly to one side, developing a pattern that resembles a one-quarter sphere. Most sidewall sprinklers are listed for light hazard, with some that have an ordinary hazard listing. Extended coverage sidewall sprinklers have larger areas of coverage than regular sidewall sprinklers. These are often used in retrofits in existing buildings. Just as for regular sidewall sprinklers, the extended coverage is listed for light hazard occupancies such as hotels.

TEMPERATURE RATINGS

Generally, sprinkler heads are tested for activation temperatures by placing them in a liquid and raising the temperature of the liquid very slowly until the sprinkler operates. Temperature ratings of fusible element–type sprinkler heads are stamped on the soldered link or color-coded on the frame arms. Temperature ranges for typical sprinkler applications are from 135 to 170°F (57 to 77°C); see table 4.4. If the ambient temperature of a location is greater than 100°F (38°C), ordinary sprinklers should not be used because the margin of safety is reduced. Higher temperature heads would need to be installed. If temperature conditions of an area are not known, then thermometers should be employed to determine the maximum ambient temperatures to assist in determination of the appropriate head temperature. Bulb-type sprinkler heads are required to have the temperature marked on the detector.

ON/OFF SPRINKLERS

Some situations may require the use of specialized sprinkler heads to reduce water damage. Libraries, for example, would not want excessive water runoff from a fire to

TABLE 4.4

Temperature Color Coding for Sprinkler Heads

Maximum Ceiling Temperature°F°C	Temperature Rating°F°C	Temperature Classification	Color Code	Glass Bulb Colors
10038	135-17057-77	Ordinary	Uncolored or black	Orange or red
15066	175-22579-107	Intermediate	White	Yellow or green
225107	250-300121-149	High	Blue	Blue
300149	325-375163-191	Extra high	Red	Purple
375191	400-475204-246	Very extra high	Green	Black
475246	500-575260-302	Ultra high	Orange	Black
625329	650343	Ultra high	Orange	Black

damage unburned collections. Central Sprinkler and Grinnell are the only companies that I know of that make the on/off type of head. They recommend that they only be used in special circumstances. This type of head has two different working components. The first works like a regular sprinkler link. Secondly, a heat-operated valve or snap disk is employed. When a fire occurs, both functioning components sense the rise in temperature from the fire and release. Following the cooling of the environment around the sprinkler, the secondary valve or snap disk detects the decline in temperature and automatically shuts off, ending the movement of water through the sprinkler head. If the fire rekindles, the secondary valve or snap disk once again opens to allow for water flow. Once they operate, on/off heads need to be replaced because the seal on the valve may leak and may not withstand surge pressures that may occur.

SPRINKLER PIPING

Sprinkler piping comes in various types and quality depending on the intended use. Pipe can be made out of copper, steel, or plastic. Schedule 40 steel is among the most durable of sprinkler piping, although other schedules of piping are permitted by the code. Plastic piping is generally limited to residential installations. When piping is exposed to corrosive atmospheres or will be installed outside, galvanized-type pipe should be used. Once installed, all sprinkler piping should be hydrostatically tested at 200 psi and should maintain that pressure for a period of not less than two hours without pressure loss. In addition, dry pipe sprinkler systems also need to have a twenty-four-hour leak test conducted. If the pressure drops more than 1½ psi in the twenty-four-hour period, it needs to be corrected. Piping requirements are listed in NFPA 13, *Standard for the Installation of Sprinkler Systems*, and NFPA 24, *Standard for Installation of Private Fire Service Mains and Their Appurtenances.*

SPARE SPRINKLER HEADS

NFPA 13 requires that every sprinkler system have a spare stock of sprinkler heads in a cabinet near the sprinkler system riser (figure 4.12). The minimum number of

FIGURE 4.12 NFPA 13 requires that every sprinkler system have a spare stock of sprinkler heads in a cabinet near the sprinkler system riser.

heads required is six (table 4.5). These sprinklers are to be available in the event that any sprinklers have operated so they can be quickly replaced. Spare sprinklers should be the same types and temperature rating as the sprinklers installed in the building. Storage temperatures for spare sprinkler heads should not exceed 100°F. A special sprinkler wrench should also be provided and located in the spare sprinkler head cabinet for the removal and reinstallation of sprinkler heads. Each type of sprinkler head should have a wrench in the cabinet. Where the sprinkler quantity in a building is larger than the minimum of six spare heads, additional heads of the proper type and temperature rating shall be provided.

STATIONARY FIRE PUMPS

Firefighters should be familiar with buildings that have fire pumps and know their locations in the building and their capacity through pre-fire plans, signs in buildings, notation on annunciator or fire alarm panel, or building information packets. Some-one from the first arriving companies should be assigned to check the fire pump room

TABLE 4.5

Stock of Spare Sprinkler Heads Required

Total Number of Sprinkler Heads	Required Number of Spare Sprinkler Heads
Under 300 heads	No less than 6 spare heads
300 to 1000 heads	No less than 12 spare heads
Over 1000 heads	No less than 24 heads

FIGURE 4.13 Someone from the first arriving companies should be assigned to check the fire pump room to make sure the fire pump is running if there is a fire in the building.

to make sure the fire pump is running if there is a fire in the building (figure 4.13). If not, the pump can be manually activated. The person assigned to the fire pump room should stay there throughout the firefighting efforts. In high-rise buildings there are usually firefighter phones installed in the fire pump room to maintain communications with the fire command center. Fire pumps should remain active until all suppression efforts are completed. Following a fire in the building where the fire pump operated, maintenance personnel should check out the fire pump to make sure it has not been damaged during the operation and that it is properly shut down and reset for automatic operation.

Fire pumps are installed in buildings when the public water main, gravity tanks, reservoirs, or other source of water pressure is not adequate or available to support the fire protection systems in a building. Fire pumps can only increase the pressure in the system; they do not increase the volume of water available to support fire suppression systems. If more water is needed, then a storage tank or reservoir is required. Fire pumps are like having a fire department engine company in the building 24/7. High-rise buildings may require fire pumps at multiple levels of the building to provide enough pressure to deliver water to the sprinklers and standpipes at the proper pressure on the upper floors. While it is unknown exactly when fire pumps were developed, it is clear that fire pumps were widely used in the early 1900s. The advent of the high-rise building made it necessary to be able to get water to the upper floors at the proper pressure to support sprinkler systems.

Fire pump installations have two basic parts, the pump and the driver. There are two primary types of fire pumps used in buildings, centrifugal split case and the vertical turbine. Centrifugal split case pumps are usually installed in situations where incoming water to the pump is under pressure. Vertical turbines are utilized where water supply is from a static source such as well, cistern, or body of water. The fire pump driver is the method by which the fire pump gets energy to operate (figure 4.14). Early fire pumps were almost exclusively steam driven. Today's modern fire pumps are electric, diesel engine, or steam turbine driven. Gasoline-, natural gas-, or propane-powered engines are not recognized by NFPA 20 as drivers for fire pumps.

FIGURE 4.14 The fire pump driver is the method by which the fire pump gets energy to operate; in this case, the driver is powered by electricity.

Utilizing gasoline, propane, or natural gas severely increases the possibility of fire associated with the driver and fuel supply. Fire pumps can be located on any floor of a building but are generally located in the basement, first floor, at the top, or other floors when more than one fire pump is required. Any type of fire pump driver will require the installation of a fire pump controller, which controls the operation of the fire pump driver (figure 4.15). Early fire pump controllers were only able to be

FIGURE 4.15 Any type of fire pump driver will require the installation of a fire pump controller, which controls the operation of the fire pump driver.

manually started. Today's fire pump controllers are completely automatic but have a manual start function as well for maintenance and testing and to manually start the pump if the automatic function fails. Controllers for electric drivers serve to provide electric current to the driver and monitor pump operations, including start and stop and pressure monitoring. Any time an electric driver is operated it must run for a minimum period of ten minutes, whether started for testing, maintenance, or by malfunction. Diesel engine controllers service to start and stop the diesel engine and other functions, similar to the electric controller. Diesel engines require a cool-down period, and when pumping operations are complete, the engine has a timer that continues the engine running for thirty minutes. NFPA 20 is the primary *Standard for Installation of Stationary Pumps for Fire Protection*. NFPA 20 requires electrical monitoring of pump controllers for pump running, power failure, or controller failure. These remote alarm signals are often incorporated into fire alarm panels or annunciators or both, so that fire departments may identify the status of a given fire pump. A fire pump controller is the enclosure that contains controls and status indicators for a fire pump. NFPA 20 requires these devices to be within sight of the fire pump motor or engine. The automatic transfer switch, which is often in a separate enclosure, transfers power to a secondary power source (when provided) in the event that primary power fails in the building or building power needs to be shut off for operational issues (figure 4.16). Fire service personnel may need access to this equipment during the course of a fire.

NFPA 20 contains reliability requirements for the power supply to an electrically driven fire pump. For example, power supply lines must be protected and the circuit must be independent of a building's electric service. The latter feature allows the fire service to shut down building power while the fire pump continues to run. OSHA electrical requirements found in Code of Federal Regulations (CFR) Subpart S must also be followed for fire pump installations. The most desirable location for a

FIGURE 4.16 The automatic transfer switch, which is often in a separate enclosure, transfers power to a secondary power source (when provided).

fire pump is in a separate building. This affords the most protection for the pump, driver, and controls from fire and gives firefighters easy access to the pump and its controllers. If locating the pump in a separate building is not possible, a fire-rated room with an outside entrance is the next best location. Recent code changes have dictated that fire pumps be installed in two-hour-rated rooms in non-sprinkled or partially sprinkled buildings and one-hour rooms in fully sprinkled buildings. However, the AHJ can require full two-hour separation even in a fully sprinkled building and this would afford the best protection for a fire pump that must be placed in a building. Rating the fire pump room is required even if the pump is within a mechanical room and helps to ensure survivability of the pump if a fire were to occur in the surrounding area. The fire pump controller as well as all of the valves and other devices associated with the fire pump should be in the rated room. Information on fire pumps is also located in NFPA standards 11, 11A, 13, 14, 15, 22, 25, and 750. Inside and outside entrances to fire pump rooms should be labeled with signage. Minimum lettering size should be six inches high with a ½-inch stroke.

Fire pumps range in size from twenty-five to five thousand gallons per minute (95 to 18,925 L/m) and greater at pressures of 40 to 394 psi (276 to 2,758 kPa) for horizontal pumps. Vertical pumps have a pressure range from 26 to 510 psi (517 to 3,448 kPa). Fire pumps are required to be tested periodically and have a test apparatus built into the system piping to allow for the test operation to take place. This apparatus is known as the fire pump test header (figure 4.17). It is different in both appearance and water flow direction from the fire department connection (FDC). The FDC has female connections, and the test header has male connections. Test headers also have shutoff valves that may or may not be attached under normal circumstances. FDC and fire pump test header connections should be clearly marked

FIGURE 4.17 Fire pumps are required to be tested periodically and have a test apparatus built into the system piping to allow for the test operation to take place. This apparatus is known as the fire pump test header.

as to their function. Test headers not only let you know that the building has a fire pump from the outside of the building but can also provide a clue as to the size of the fire pump. Each connection on the test header represents 250 gpm of fire pump capacity. Multiply the number of connections by 250 and you know the size of the fire pump. Fire pump information is annunciated at the fire alarm control panel and at the annunciator panel. When firefighters arrive at an alarm in a building, they should check the annunciator or fire alarm control panel first to determine the origin of the alarm and should also make note of the fire pump status. If the fire pump is running, in addition to a water flow switch, it may be more of an indication of actual sprinkler head activation. It is possible, but unlikely, that both a flow switch and the fire pump would malfunction at the same time.

In a major city on the East Coast, an arson fire recently occurred in a high-rise building. The fire department arrived with their usual one engine and one ladder response for alarm bells sounding. The annunciator panel indicated that there was a water flow alarm on the second floor of the building. It is unclear whether the firefighters also noticed that the fire pump was running. However, they checked the second floor, found nothing, and left. When the building plumbing staff and fire marshal arrived, they noticed that the fire pump was running in addition to the flow alarm. They once again checked the second floor and found nothing. Then they started at the top of the eight-story building and searched each floor. When they got to the fifth floor, the corridor was full of water. The water was traced to an activated sprinkler head in a room on the floor where a fire had been set, which caused the sprinkler to activate and the fire pump to run. This incident is a good example of what happens with frequent false fire alarms in buildings in a jurisdiction. Firefighters become complacent about fire alarm response and do not do a thorough job of checking out the cause of the alarm. In this case there was significantly more water flow and cleanup required than if the firefighters had taken note of the flow and fire pump running notifications on the panel and searched the entire building, as the building staff ended up doing. When firefighters determine that a fire pump is operating in a building, they should notify or locate the building maintenance personnel. These personnel should go to the fire pump room or location and place the fire pump in manual operation and the pump should run until the emergency is over. When the pump is no longer needed, it can be shut down manually, serviced, and restored to automatic operation.

FIRE PUMP TRIM

In addition to the fire pump, the system must also be fitted with several auxiliary appurtenances, including relief valves, hose valves, automatic air release valves, and circulation relief valves. All of these devices are located within the rated fire pump room. Relief valves are installed when pump pressures may exceed the pressure rating of the fire protection system (figure 4.18). Automatic air release devices should be installed on the top of pump casings where the casing is normally filled with water. Pumps that start automatically must also be equipped with circulation relief valves. Circulation relief valves open slightly to allow for water to flow, preventing overheating of the pump when water is not flowing in the system (churn; figure 4.19). Circulation relief valves are only installed on electric motor–driven pumps.

FIGURE 4.18 Relief valves are installed when pump pressures may exceed the pressure rating of the fire protection system.

FIRE PUMP CONTROLLER

Electric fire pump motors require the installation of a controller to ensure the proper automatic operation of the fire pump (figure 4.20). Controllers are large cabinets with levers and indicator lights on the outside to turn the pump power off and on manually and determine the status of the pump. The controller cabinet is generally located in close proximity to the fire pump and within the fire pump room when

FIGURE 4.19 Circulation relief valves open slightly to allow for water to flow, preventing overheating of the pump when water is not flowing in the system (churn).

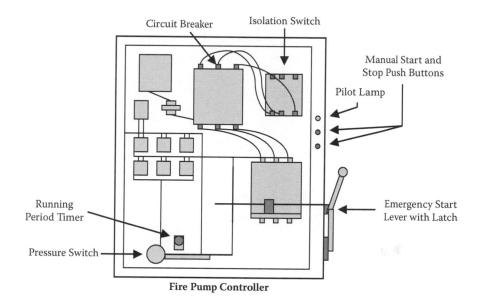

FIGURE 4.20 Electric fire pump motors require the installation of a controller to ensure the proper automatic operation of the fire pump.

there is one. Most controllers are painted red for easy identification and to distinguish them from other electrical equipment that may be located in a mechanical area. Electric fire pump controllers contain disconnecting means, timers, circuit breakers, and other associated equipment within the cabinet. Controllers are listed devices and completely assembled and tested at the factory. They are ready to attach to the power supply for the fire pump and driver motor. Engine-driven fire pumps also require the installation of a controller. Controllers, installation, and purpose are much the same for the engine-driven fire pump as for the electric motor–driven fire pump. Engine-driven pumps require additional monitoring for low oil pressure, water temperature, engine start failure, and over-speed shutdown.

Jockey Pump

Piping, size of systems, leaks, and fluctuations in public water supplies can cause changes in levels of water within fire protection system piping. In order to prevent the fire pump from constantly starting and stopping, to maintain the water in the system jockey pumps are installed (figure 4.21). Every fire pump should be equipped with a jockey pump to maintain the water levels without causing the fire pump to start up. While jockey pumps can maintain water levels and pressure on the system, they do not provide enough capacity to support the activation of a single sprinkler head on the sprinkler system. When a sprinkler head is activated by a fire, the fire pump will kick in and provide water and pressure to the sprinklers. The pump will continue in operation until shut off by firefighters or maintenance personnel. If the floor zone valve is shut off to change a sprinkler head once the fire is out, maintenance

FIGURE 4.21 In order to prevent the fire pump from constantly starting and stopping, to maintain the water in the system, jockey pumps are installed.

personnel should also shut down the fire pump so that it will not overheat. When the sprinkler heads are replaced, the jockey pump will fill up the piping on the line where the head or heads were changed out. Once the system is full of water again, the fire pump should be placed back in the automatic mode and the fire alarm system reset.

Sprinkler heads are not the only fire protection devices that have been subject to recalls. Approximately 18,300 jockey pumps manufactured by Water Technology, Inc., of Auburn, New York, were recalled by order of the Consumer Product Safety Commission (CPSC) in 2006. The pumps were sold under the Goulds Pumps®, Bell & Gossett®, and Red Jacket Water Products brands. Models NPE, NPO, MCC, MCS, SM, or Series 3530 are marked on the name plates. Recalled pumps were manufactured between December 2005 and July 2006. For more information, call 800-984-9199 between 8 a.m. and 5 p.m. ET Monday through Friday or visit the Website at www.goulds.com or www.bellgossett.com. To view this recall on CPSC's Website, including photographs of the recalled products, go to: http://www.cpsc.gov/cpscpub/prerel/prhtml06/06233.html

RESIDENTIAL SPRINKLERS

The United States has the highest fire death rate of the industrialized world, 13.4 people per million. Each year, over the past several decades, 2,500 to 4,000 people have died in residential fires (over 80 percent of all fire deaths occur in the home). This is a marked reduction from the 8,000-plus people who died prior to the Fire Prevention & Control Commission report *America Burning*, which was released in 1973. However, in spite of all the efforts by the NFPA and United States Fire Administration (USFA), too many people are still dying in residential fires. Many of those fatalities are elderly (over one hundred people per million) and children (over forty per million). The home is by far the most dangerous place to be in terms of fire

FIGURE 4.22 The home is by far the most dangerous place to be in terms of fire deaths and injuries.

exposure (figure 4.22; table 4.6). By far the most common area of origin in home fires (30.9 percent) is the kitchen, generally caused by unattended cooking operations or human error (figure 4.23). Bedrooms are the second most frequent location at 13.6 percent, followed by the living/family room areas at 8.0 percent. The USFA has proposed a stopgap measure of just putting sprinkler heads in kitchens to combat fires that occur in the most common area of origin. While the kitchen is the leading area of origin and the location of the most fire injuries, it is the bedroom where the most fire deaths occur. The leading cause of fire deaths in the United States is careless smoking, followed by arson and heating equipment.

Smoke alarms have come a long way toward reducing fire deaths where they are installed, but 39 percent of the fires and 52 percent of the fire deaths occur in homes without smoke alarms. However, the real answer lies in a combination technology of properly tested and maintained smoke alarms and residential sprinkler systems. Unfortunately, as previously mentioned, there is significant resistance to

TABLE 4.6
Room of Origin of Residential Fires

Room of Origin	Number of Fires	Percentage of Total Fires
Kitchen	125,600	30.9
Bedroom	55,300	13.6
Living room/family room	32,700	8.0
Chimney	25,900	6.4
Laundry room	18,900	3.7
Exterior	13,200	3.2
Garage or vehicle storage	12,700	3.1

FIGURE 4.23 The most common area of fire origin in home fires (30.9 percent) is the kitchen, generally caused by unattended cooking operations or human error.

the installation of residential sprinklers in many parts of the country. Contractors resist because of the increased cost they fear. Politicians resist because they do not understand the importance of residential sprinklers and bow to the pressures of contractors and other special interest groups. Fire chiefs and fire prevention personnel resist because of the pressure applied by politicians. It's a vicious circle that needs to be broken once and for all. It takes courage to pass residential sprinkler legislation. Montgomery County, Maryland, recently passed and the county executive signed into law a residential sprinkler bill requiring sprinklers in all new home construction. That is a step in the right direction. But we need to take it further. Existing homes may be around for one hundred years or longer. There need to be retroactive requirements phased in over a reasonable period of time to require sprinkler systems in all residential occupancies.

In an effort to combat the serious loss of life in homes, the residential sprinkler was developed to go with the residential-type sprinkler system. NFPA has established separate sprinkler standards for residential sprinklers, NFPA 13R and 13D. NFPA 13R is the *Standard for Installation of Sprinkler Systems in Residential Occupancies up to and Including Four Stories in Height.* NFPA 13 D is the *Standard for Installation of Sprinkler Systems in One- and Two-Family Dwellings and Manufactured Homes* (table 4.7). The residential sprinkler code allows for some installation types of materials not permitted in NFPA 13, such as the use of listed plastic pipe, use of combination domestic water piping and sprinkler piping to add sprinklers to dwellings, and the omission of sprinklers in closets no larger than 24 ft² and bathrooms no larger than 55 ft². Sprinklers are also allowed to be omitted from garages, carports, attics, and crawl spaces not intended for living spaces and not used for storage. Entrance foyers, which are not the only means of egress, do not have to be sprinkled. Special heads were also developed for use in residential sprinkler systems to allow for quick response of sprinklers compared to conventional sprinkler

---

TABLE 4.7
Design Criteria for Residential Sprinkler Systems (NFPA 13D)

Design	Criteria
Minimum water flow	18 gpm
Maximum single sprinkler coverage area	144 ft^2
Maximum distance between sprinklers	12 ft
Minimum distance between sprinklers	8 ft
Maximum distance sprinkler from wall	6 ft

---

heads in nonresidential settings. Factory Mutual in association with the USFA led the way in residential sprinkler development.

RECALLED SPRINKLER HEADS

On October 14, 1998, the Consumer Product Safety Commission in conjunction with Central Sprinkler Corporation announced the mandatory recall of the Omega quick response sprinkler heads (figure 4.24). Originally tested and approved for installation, it was found that some of the Omega heads, after being in service for a period of time, did not function properly during actual fire incidents. The problem occurred from the use of oils and solvents during installation of the sprinkler piping that reacted with the rubber O-ring in the sprinkler head. The chemical contamination caused the O-rings to swell, preventing the proper activation of the sprinkler head during fires. Heads were removed from systems across the country and tested to determine the scope of the problem. It was found that 30 percent of the tested sprinkler heads failed the test and did not function. All 8.4 million wet and dry sprinkler heads manufactured between 1983 and August 1998 were recalled (table 4.8; table 4.9). Central Sprinkler Company provided replacement heads for the recalled

FIGURE 4.24 Central Sprinkler Corporation announced the mandatory recall of the Omega quick response sprinkler head during October 1998.

TABLE 4.8

Affected Models Central "Wet" Sprinklers (Manufactured from 1989 to 2000)

GB	GB4-FR	GB-R1	BB2	ELOC	ELO-GB QR
GB-J	GB4-EC	GB-RS	BB3	ESLO	LD
GB-1	GB4-QREC	GB-R	SD1	ELO SW-20	K17-231
GB-ALPHA	GB-20	ROC	SD2	ELO SW-24	Ultra K17
GB4	GB-20 QR	BB1 17/32	SD3	ESLO-20GB	ELO-16 GB
GB-QR	GB-LO	BB2 17/32	HIP	ELO-231GB	GB Multi-level
GBR-2	LF	BB3 17/32	WS	ELO-GB	GB-QR Multi-level
GB-EC	GBR	BB1	ELO-LH	ELO-231 GBQR	ELO-16 GB FR

heads to owners who applied for the heads during the application period, which has since ended. There may, however, still be Omega sprinkler heads installed, and even though the replacement program is over, they should still be replaced.

Central Sprinkler Company, which is a subsidiary of TYCO Corporation, announced the recall of an additional 35 million sprinkler heads, including seventy-seven different models due to potential O-ring problems. This recall followed an earlier recall of the 8.4 million Omega sprinkler heads. In addition to the Central heads, approximately 167,000 O-ring heads manufactured by Gem and Star were also part of the recall (table 4.10). These heads were manufactured from the mid-1970s to 2001.

During August 1999, the CPSC announced a recall of Star sprinklers manufactured by the Mealane Corporation of Philadelphia (figure 4.25). Up to one million sprinkler heads were voluntarily recalled that were manufactured between 1961 and 1976 (table 4.11). These sprinklers could fail during a fire. Over 67 percent of the heads submitted for testing by independent testing laboratories failed to activate. Models affected are the D-1, RD-1, RE-1, E-1, and the ME-1. All of these heads were used in dry pipe sprinkler systems. The former Star Sprinkler Company sold its assets in 1976 and became known as Mealane Corporation.

Firematic Sprinkler Devices has announced that Underwriters Laboratory (UL) is investigating the Firematic nickel-plated model S&A 160° sprinkler head. Testing of a single sample from a single location indicated that some of the heads did

TABLE 4.9

Central "Dry" Sprinklers (Manufactured From Mid-1970s to 2001)

A-1	GB	GB4-EC	ELO-16 GB
H-1	GB-QR	GB4-QREC	ELO-16 GB`
J	GB4	ELO-231 GB	
K	GB4-FR	ELO-GB QR	

TABLE 4.10

GEM "Wet" Sprinklers (Sold under Gem name From 1995 to 2001)

F927

not operate as intended. UL is requesting that samples of Firematic S&A heads be removed from each installation and submitted for testing to determine if this problem is more widespread. These sprinklers were manufactured between 1976 and 1979. These sprinklers are equipped with a heat-responsive element made of solder that melts at a given temperature. Once the solder melts, the heat-responsive element is supposed to separate from the sprinkler body to allow the water to flow from the head. The nickel plating on the heat-responsive element may prevent solder from melting properly and thus preventing the flow of water from the sprinkler head. During a fire, this feature may prevent the activation of the sprinkler heads. UL has requested that additional samples of this type of sprinkler head be submitted for testing. Models affected include the "D," TU-57, TP-57, TU-80, TU-29, TP-29, TU-39, and TP-39. Temperature classifications range from 160, 212, 286, and 360°F. All of the models and temperature classifications of heads should be tested by random sample. NFPA 13 gives guidance of sampling of installed sprinkler heads. As a direct result of the recent sprinkler recalls, the NFPA has implemented the mandatory testing of random samples of sprinkler heads at specific intervals with the 2002 version of NFPA 13. This was done in an attempt to ensure that any problems with sprinkler heads are discovered and addressed before fires occur.

During the late 1990s, the University of Maryland, Baltimore, started experiencing spontaneous disintegration of Grinnell dry pendent sprinkler heads that were

FIGURE 4.25 During August 1999, the CPSC announced a recall of Star sprinklers manufactured by the Mealane Corporation of Philadelphia.

TABLE 4.11
Star "Dry" Sprinklers (Manufactured From 1996 to 1998)

ME-1	SG
SG-QR	Q
Q-QR	

used in an underground parking garage. The heads would come apart and water would flow without any outside intervention. Samples were sent to Grinnell and they ended up replacing all of the heads in the six-level parking garage without cost to the university. The problem seemed to be with the environment in the parking garage and not any problem with sprinkler heads, so no head recall was conducted.

SPECIALIZED SUPPRESSION SYSTEMS

Alternative extinguishing systems are often employed where automatic fire suppression is required or preferred and water will cause damage or other type of undesirable result when discharged into the space. Wherever possible, water sprinkler systems should be used for fire protection purposes. There are, however, some applications that, because of fuel type, process, or equipment sensitivity, require the use of specialized suppression systems (figure 4.26). These include gases such as carbon dioxide, halon, nitrogen, and halon replacements. Dry chemical systems

FIGURE 4.26 There are some suppression applications that, because of fuel type or process or equipment sensitivity, require the use of specialized suppression systems that do not use water.

may also be used for certain applications such as cooking operations and flammable liquid hazards. Foam systems may also be used as suppression systems for flammable hazardous materials. Regardless of the type of suppression system, firefighters should be familiar with the systems, operation, and how to deal with fires when these systems are in place. Suppression systems are designed for the hazard they are protecting and outside intervention by responding firefighters before a fire is extinguished may make the situation worse. For example, halon and replacement gases are engineered to provide a certain concentration of the extinguishing agent in a room to put out the fire. If responding firefighters open the room prematurely, the extinguishing gas could escape and the fire may rekindle or get worse. Firefighters should also understand the dangers of oxygen deficiency created by some gases such as carbon dioxide, which require the use of self-contained breathing apparatus (SCBA) even when there is no visible smoke or gas. Suppression systems, if properly installed and maintained, should extinguish any fire that occurs, and by the time the firefighters arrive, the fire should be out and all that needs to be done is overhaul and putting the system back in service.

For over fifty years, from 1910 to the late 1960s, when a clean, dry, gaseous fire-extinguishing agent was required, carbon dioxide was the choice. It was in fact the only gaseous fire-extinguishing agent available. In the late 1960s, the halons became commercially viable alternatives to carbon dioxide for applications where the life safety risks posed by carbon dioxide were considered unacceptable. In particular, halon 1301 became popular as a fire protection agent for total flooding spaces where personnel might be present. For over twenty years, Halon 1301 was the fire suppression agent of choice for high-value hazards that required a clean, "life safe" agent. However, it has now been determined to harm the Earth's ozone layer, and manufacturer of the agent has been banned in the United States and other countries by treaty. Some of these alternative agents can have adverse environmental impacts, such as contaminating groundwater or depleting the ozone layer above the Earth.

Carbon dioxide systems are installed according to NFPA 12, *Carbon Dioxide Extinguishing Systems*. Carbon dioxide (CO_2) is a colorless and odorless, nonflammable gas. It is heavier than air and will tend to "hover" above the surfaces of flammable liquids. CO_2 does not react chemically with most substances and provides its own pressure discharge from its container. It has been used as an extinguishing agent in suppression systems and portable fire extinguishers for many years. Carbon dioxide is designed to extinguish flammable liquid and electrical fires. It does not conduct electricity, which makes it safe for electrical equipment and components. CO_2 works well for sensitive processes and electrical equipment because it does not leave a lasting residue once discharged, so it does not damage equipment. When discharged, CO_2 forms a finely divided solid material, sometimes referred to as "snow" or "dry ice." This material quickly melts at elevated temperatures in the atmosphere, leaving no residue behind. When discharged, carbon dioxide also will leave a frost on piping and other component parts that can cause tissue damage if contacted with bare skin. During the discharge of CO_2, static electricity is also formed and can cause electric shock to firefighters or become an ignition source in an explosive atmosphere if the equipment nozzles are not properly grounded. Carbon dioxide works by excluding oxygen from the fire and the fire is smothered.

It does not have any significant cooling action, so if there are hot surfaces and fuel present, the fire may rekindle when oxygen is reintroduced into the area. Carbon dioxide is not very effective against ordinary combustible materials because it does not cool them down below their ignition temperature. Because carbon dioxide displaces the oxygen in the air to extinguish the fire, it also presents an asphyxiation hazard to firefighters. There may not be enough oxygen in the room where CO_2 was discharged to support life, and firefighters will need to wear SCBA. Both audible and visual predischarge alarms are required for carbon dioxide suppression systems to allow occupants to evacuate the area safely before the discharge of carbon dioxide into the area occurs.

Carbon dioxide suppression systems can be installed as total flooding, local application, or using hand hose lines. No matter which system is used, the suppression ability is restricted by the amount of agent available in tanks attached to the system. Once the tank is empty, suppression slows or is stopped.

Halogenated agents (halons) were developed in the late 1800s with the first of the agents carbon tetrachloride (Halon 104). Methyl bromide (Halon 1001) was discovered in the late 1920s. It is highly toxic and was used in aircraft and ships during World War II. Bromochloromethane (Halon 1011) was developed by the Germans during World War II to replace methyl bromide. Underwriters Laboratory began tests on halon agents in 1947 and found that carbon tetrachloride and bromochloromethane had comparable toxicity, but the latter was a better extinguishing agent. Increased use of dry chemicals as extinguishing agents in the 1950s brought an end to the early halons (104, 1001, and 1011). Purdue University conducted tests on new fire-extinguishing agents in 1947 and narrowed down sixty potential agents to four: dibromodifluoromethane (Halon 1202), bromochlorodifluoromethane (Halon 1211), bromotrifluoromethane (Halon 1301), and dibromotetrafluoroethane (Halon 2402). Halon 1202 was determined to be the best extinguishing agent but also was the most toxic. Halon 1301 was the second best extinguishing agent and the least toxic. The military was experiencing an alarming number of fires in various locations on ships and land and needed an extinguishing agent to deal with the problem. Halon 1301 emerged as the agent of choice for the military and commercial applications in the United States. Halon 1211 was widely used in Europe. However, it was soon determined that Halon 1211 was not a suitable total flooding agent where people were exposed during discharge, and it was abandoned for those situations. Halons are expensive but very effective extinguishing agents that have been widely used until the present time. NFPA Standards 12A and 12B were adopted for Halon 1301 and 1211. During the 1990s, it was discovered that halons were suspected of causing damage to the Earth's ozone layer. During 1994, the production of halon agents started to be phased out. Many systems still remain in place, but a transition to other fire suppression agents is underway.

Halons can be liquids or gases. They are nonflammable and composed of combinations of carbon, fluorine, chlorine, and bromine. Halon 1301, 1211, and 2402 are not corrosive to modern construction materials. This is a characteristic that makes them good extinguishing agents for sensitive computer and electronics equipment. Fluorine in a halon compound serves to stabilize the compound, reduce toxicity, reduce boiling point, and increase thermal stability. The presence of chlorine provides fire extinguishment effectiveness, increases boiling point, increases toxicity, and reduces thermal stability. Bromine in a halon compound serves to provide the same effects as chlorine but to a greater degree. Halon compounds containing fluorine, chlorine, and bromine can possess varying degrees of extinguishing effectiveness, chemical and thermal stability, and toxicity. Numbers in a halon compound name actually have meaning to them in terms of the chemical makeup. The first number indicates the number of carbon atoms in the compound. The second number indicates the number of fluorine atoms in the compound. The third number indicates the number of chlorine atoms, and the fourth number indicates the number of bromine atoms. Therefore, halon 1301 has one atom of carbon, three atoms of fluorine, zero atoms of chlorine, and one atom of bromine.

Generally, the toxicity of halon extinguishing agents is low if the exposure time is short. Installation of systems does not require the same type of predischarge alarm that is required for carbon dioxide systems. Exposure times for halon systems have been established by the percentage of volume of agent in an occupied enclosure. Halon 1301 is the least toxic of the halons. At concentrations up to 7 percent of the total volume of air, persons can be safely exposed for 15 minutes maximum. For 7 to 10 percent concentrations, one minute is the maximum. From 10 to 15 minutes, the exposure concentration is thirty seconds maximum. At concentrations of halon 1301 above 15 percent concentration, exposure should be avoided. Normally occupied areas are permitted by code to have design concentrations of halon 1301 up to 10 percent. Fifteen-percent concentrations are permitted in areas that are not normally occupied. Because the toxicity levels of halon 1211 are below the levels necessary for extinguishment, systems using 1211 are not permitted in normally occupied areas. The bottom line is that occupants should evacuate quickly and firefighters should wear SCBA when entering areas where halon agents have been discharged.

While halon can no longer be manufactured in the United States, there are no requirements that halon systems still in use be replaced with clean agents. There are also supplies of halon still available that can be used to refill existing systems, which will remain in place until discharged or requirements for halon use in existing installations changes.

Numbers for halon agents are based upon the chemicals present in the formulation, which includes carbon and the first four members of Family VII on the Periodic Table of the Elements. Family VII is composed of fluorine, chlorine, bromine, and iodine. The first digit of a halon number is the number of carbon atoms in the formulation. Second is the number of fluorine atoms and the third digit is the number of chlorine atoms. The final digit is the number of bromine atoms. If a fifth number is present, it represents the number of iodine atoms. Using that information Halon 1301 would have one atom of fluorine, three atoms of chlorine, no atoms of bromine, and one atom of iodine (figure 4.27).

CLEAN AGENTS

Because of the stratospheric ozone-depleting characteristics of halon agents, replacements for halon have been developed. Clean agents are defined as those that vaporize readily and leave no residue. NFPA 2001 is the *Standard on Clean Agent Fire*

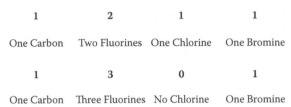

Naming of Halons

1	2	1	1
One Carbon	Two Fluorines	One Chlorine	One Bromine

1	3	0	1
One Carbon	Three Fluorines	No Chlorine	One Bromine

FIGURE 4.27 Numbering system for the naming of halon extinguishing agents.

Extinguishing Systems. Clean agents are chemical fire-extinguishing agents developed in response to international restrictions on the production of certain halon fire-extinguishing agents under the Montreal Protocol signed September 16, 1987, as amended. Clean agents are total flooding fire-extinguishing agents. If the fuel is electrical equipment (Class C fuels), such as computer or telecommunications equipment, virtually all of the clean agents in NFPA 2001 are listed and approved for such hazards when electric power to the equipment is shut down upon discharge of the clean agent, which changes the fire to Class A fuels once the electrical equipment is deenergized. NFPA 2001 recognizes two levels of protection for flammable liquids and gases: flame extinguishment and creating an inert atmosphere in the space. NFPA 2001 requires inerting concentrations to be used for hazards involving flammable liquids and gases (Class B fuels), providing that conditions for subsequent flashback or explosion may exist. Annex A of NFPA 2001 describes such conditions. Practically speaking, nearly all hazards where flammable liquids are the primary fuel load would best be protected by systems employing an inerting concentration.

Two general classes of clean agents can be identified, halocarbon compounds and inert (not normally chemically reactive) gases and mixtures. Halocarbon compounds are made up of carbon, hydrogen, fluorine, chlorine, bromine, and iodine. They can be grouped into five general categories, hydrobromofluorocarbons (HBFC), hydrofluorocarbons (HFC), hydrochlorofluorocarbons (HCFC), perfluorocarbons (FC or PFC), and fluoroiodocarbons (FIC). As a group, these agents are all liquefied gases, therefore requiring less storage space than the inert gaseous halon replacements that are just gases. Liquefied gases have high expansion ratios and you can transport and store more of a gas that has been liquefied than you can a non-liquefied gas. They are nonconductive, total flooding gases, decompose primarily to (HF), are less effective as extinguishing agents than halon 1301, and are more expensive. These agents have varying levels of toxicity and manufacturers' material safety data sheets (MSDSs) should be consulted for toxicity information. As a practice, firefighters entering areas where agents have been discharged should wear full turnouts and SCBA for protection. Inert gas halon replacements include nitrogen and argon. There is also one agent that uses a very small amount of carbon dioxide. Inert agents are all pressurized gases and require larger storage containers than do liquefied gases. As a group they are nonconductive. Inert gas replacements are nontoxic; however, they can cause asphyxiation resulting from displacement of oxygen in the air. Firefighters should wear full turnouts and SCBA when entering areas where extinguishing agent discharge has occurred.

There are currently several commonly used clean agent replacements for halons, including Intergen (IG-541), FM200 (HFC-227ea), Agrotec (IG-01), and HFC-23 (FE13). Intergen is a blend of nitrogen, argon, and carbon dioxide and acts by reducing the oxygen level of a space, making it inert to combustion. FM 200, with a chemical name of heptafluoropropane, CF_3CHF_2, acts by inhibiting combustion. Agrotec (IG-01) is the inert gas argon and acts by reducing oxygen. HFC-23 (FE13), trifluoromethane or CHF_3, acts by inhibiting combustion.

Dry chemical extinguishing agents and systems were developed in the early 1950s (figure 4.27). NFPA published its first edition of the standard for *Dry Chemical Extinguishing Systems*, NFPA 17, in 1952. Many of the dry chemical extinguishing agents were developed by the navy for shipboard firefighting and were eventually developed into portable fire extinguishers and suppression systems in the private sector. Dry chemical extinguishing agents should not be confused with dry powder, which is used for Class "D" combustible metal fire extinguishment. Dry chemical agents can be organized into five general categories:

- Borax and sodium bicarbonate
- Sodium bicarbonate
- Urea-potassium bicarbonate
- Multipurpose (monoammonium phosphate base)
- Purple "K" (potassium bicarbonate)

Borax and sodium bicarbonate agents were among the first dry chemical extinguishing agents developed. Sodium bicarbonate rose to the surface as the preferred agent based upon its effectiveness in extinguishing fires. Urea-potassium bicarbonate-based extinguishing agents were developed in the late 1960s in Great Britain. Monoammonium phosphate and potassium bicarbonate extinguishing agents became the next agents developed and remain in the forefront of today's modern fire extinguishers and suppression systems. Monoammonium phosphate is the multipurpose fire-extinguishing agent used in most fire extinguishers around the world. It is a yellow powder with a faint ammonia odor. Like all other dry chemical agents, it is nontoxic, but caution should be exercised around people with existing respiratory conditions. Potassium bicarbonate is one of the most effective dry chemical extinguishing agents ever developed. However, it is also one of the most expensive and its use is reserved for aircraft firefighting and other expensive equipment protection. Sodium bicarbonate has become a more specialized agent used largely in fire suppression systems for commercial cooking operations. It is generally accepted that dry chemical agents extinguish fire by interrupting the chemical chain reaction, although there is also some smothering and cooling that contributes in a small way.

Dry chemical fixed suppression systems are primarily used for flammable liquid and gas fires. There are two basic types of suppression systems, total flooding and hand hose/local application. Dry chemical agents are usually expelled from their container by gaseous nitrogen. Fixed suppression systems are automatic and operate when a sensing device in the protected area detects elevated heat and activates and shuts down process

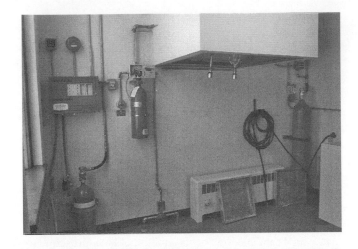

FIGURE 4.28 Dry chemical extinguishing agents and systems were developed in the early 1950s.

equipment as required by NFPA 17. Fixed suppression systems also have manual actuation devices similar to manual fire alarm pull stations to instantly activate the system (figure 4.29). When dry chemical systems are used to protect commercial cooking operations, they must also shut off the fuel supply when the system discharges.

WET CHEMICAL SYSTEMS

One of the drawbacks to using a dry chemical extinguishing system is the mess it creates in the protected area when it discharges. Cleanup and downtime for a business or operation following a discharge can be extended with dry chemical agents.

FIGURE 4.29 Fixed suppression systems also have manual actuation devices similar to manual fire alarm pull stations to instantly activate the system.

TABLE 4.12
Foam Type Comparison

	Drain Time	Viscosity	Flow Rate	Polar Solvent Compatible
Protein	Long	High	Low	No
Fluoro-protein	Moderate	Low	High	No
AFFF	Short	Low	High	No
AFFF Polar	Very short	Low	High	Yes
FFFP	Moderate	Low	High	Yes

Wet chemical systems have largely replaced dry chemical systems for commercial cooking equipment in most new installations. They are very effective and do not require any extensive cleanup like dry chemical does. Organic or inorganic salts are mixed with water, forming an alkaline solution that is expelled by an inert gas through specially designed piping and nozzles. Exact formulations will vary from manufacturer to manufacturer, so the MSDS is the best source of information about the physical and chemical properties of wet agents. Generally, wet agents are considered to be nontoxic. Because wet agents contain water, they should not be used in applications where electrical equipment or water-reactive materials are exposed to the discharge of the wet chemical. When the wet chemical contacts the fat in a grease fire, it hydrolyzes to form a foam coating on the surface, which smothers the flames and prevents reignition by blanketing the surface of the grease and excluding oxygen. There is some cooling effect involved in the extinguishing process that also occurs. Wet systems are activated in much the same way as dry systems, automatically or

TABLE 4.13

Sprinklers Designed to Control a Fire

Standard spray pendent	Standard spray upright
Large drop	Old-style/conventional
Quick response extended coverage	Standard response extended coverage
Horizontal sidewall	Vertical sidewall
Residential	Flush, recessed, concealed
Flow control (on-off)	Corrosion-resistant
High-temperature, wax-coated	Intermediate level (water shielded)
Nozzles	Open
Quick response	Special
Specific application	Spray

Sprinklers Designed to Extinguish a Fire

Early suppression, fast response	Quick response, early suppression

manually. NFPA 17A, *Standard for Wet Chemical Extinguishing Systems*, requires
fuel source shutdown before wet systems discharge.

WATER SPRAY FIXED SYSTEMS

LOW-, MEDIUM-, AND HIGH-EXPANSION FOAM SYSTEMS

Foam used for suppression systems is a mixture of air, water, and foam concentrate.
Foams are generally used to extinguish flammable liquid fires. Firefighting foam is
divided into three general categories, low-expansion 20:1, medium-expansion
20 to 200:1, and high expansion 200 to 1000:1. Foams are further classified by their
ability to extinguish hydrocarbon or polar solvent fires. There are also four general
types of foam, as shown in table 4.12. Low-expansion foams are used for fires involving flammable or combustible liquids within tanks. Application can be either surface
or subsurface. Medium-expansion and high-expansion foams are generally used
for Class A structural types of fires in difficult-to-reach areas such as basements.
Through fixed suppression systems, foam can be discharged through air-aspirating
deluge sprinklers or through regular sprinkler heads. Foam can be expelled through
an open deluge system or a closed system similar to a water sprinkler system.

5 Standpipes/FDC/ Water Supplies

INTRODUCTION TO STANDPIPES AND WATER SUPPLIES

Fire protection systems cannot function effectively nor can firefighters fight fire without an adequate water supply. Several NFPA standards and model building codes provide requirements for water supplies to buildings as well as water mains, fire department connections, and fire hydrants. These include the *International Building Code* (IBC); NFPA 5000, *Building Construction and Safety Code*; the *International Fire Code* (IFC); NFPA 1, the *Uniform Fire Code*; NFPA 101, the *Life Safety Code*; NFPA 1142, *Standard for Water Supplies for Suburban and Rural Firefighting*; NFPA 291, *Standard for Fire Flow Testing and Marking of Hydrants*; NFPA 24, *Standard for the Installation of Private Fire Service Mains and Their Appurtenances*; NFPA 22, *Standard for Water Tanks for Private Fire Protection*; NFPA 13, *Standard for the Installation of Sprinkler Systems*; 13R, *Standard for the Installation of Sprinkler Systems in Residential Occupancies up to and Including Four Stories in Height*; NFPA 14, *Standard for the Installation of Standpipe and Hose Systems*; NFPA 15, *Standard for Water Spray Fixed Systems for Fire Protection*; NFPA 16, *Standard for the Installation of Foam-Water Sprinkler and Foam-Water Spray Systems*; NFPA 750, *Standard on Water Mist Fire Protection Systems*; NFPA 1963, the *Standard for Fire Hose Connections*; and NFPA 170, *Standard for Fire Safety Symbols*. Water supplies can be both public, through the local water department, and private, supplied by storage systems or water supplies at a facility. Water supplies for fighting fires most often comes from fire hydrants supplied from water mains hooked to the local water system. Many firefighters are familiar with the slang used for connecting to a fire hydrant. It is often referred to as catching a "plug." Fire hydrants are sometimes referred to as "fireplugs." Did you ever wonder where the term "fireplug" orginated? After all, it does not seem to make sense that you get water flow from a plug! Early water mains were made of wood in the 1700s and 1800s. Hollowed-out wooden logs were used to transport water through the community (figure 5.1) to meet the daily needs of its residents. Firefighters realized that the wooden water mains could be a source of water for fighting fire. When a fire occurred, firefighters would dig down to the wooden main and drill a hole through it. Water from the pipe would then fill the excavated area, providing a drafting pit for pumps or a source of water for bucket brigades. Once the fire was extinguished, firefighters would drive a wooden plug into the drilled hole. The spot of the plug would be noted so a new hole would not have to be drilled and the water source would be available for any future fires. In larger cities, wood plugs were installed in the wooden mains as they were constructed. Firefighters would then know ahead of time where the plugs were located. Thus the term fire plug, which carries over today as a term meaning fire hydrant (figure 5.2).

157

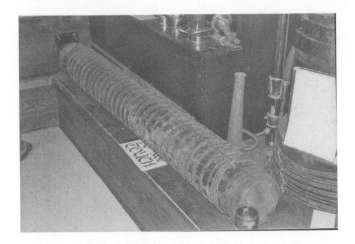

FIGURE 5.1 Hollowed out wooden logs were used to transport water through the community and were also accessed by firefighters to fight fires.

When fire protection is contemplated for a new construction project or a renovation, methods must be developed to get the proper water supply to the building fire protection systems through public or private water supplies, mains, and hydrants. Water must also be provided inside a building for fire department use when the height or size of the building is beyond the reach of fire department hoses at the street level. NFPA 101, the *Life Safety Code*, requires fire department hose valves when called for in the individual occupancy chapters (figure 5.3). Even when buildings are equipped with standpipe hose connections and sprinkler systems, fire companies

FIGURE 5.2 Early firefighters drilled holes into the wooden water mains to get water for fires and when finished would plug the hole with a wooden plug; thus today's slang reference to fire hydrants as "fire plugs."

FIGURE 5.3 NFPA 101, the *Life Safety Code*, requires fire department hose valves to be installed when called for in the individual occupancy chapters.

may need to supply additional water to a building as fire operations commence to support sprinkler systems and to supply additional water and pressure to the fire department hose valves. All issues concerning water supplies, sprinkler systems, standpipe hose valves, and hydrant locations should be considered during the plans review process in any building under construction or renovation.

WATER MAINS

Public water mains are the most common type of water supply for fire protection needs with both building fire protection systems and firefighting by the fire department. Water mains deliver varying amounts of water through varying size pipes to the familiar hydrant we see on the typical street corner. It is important that firefighters know how the public mains are laid out in a community, the size of the mains, and the volume of water available for firefighting purposes. Looped grid patterns should be used when installing new water mains. This type of system provides water from two different directions. Installing water mains in this configuration generally allows for an increase in available water, and more fire department pumpers will be able to draw water from the same main. Dead-end grids that exist in some existing locations will limit the water flow and restrict the number of pumpers that can access the mains for firefighting water supply. It is not practical and unlikely that firefighters will remember the flow characteristics of each fire hydrant in their response areas. The National Fire Protection Association developed a system of color-coding hydrants according to their fire flow to assist firefighters in determining how much water is available from any given hydrant. Hydrants can be marked according to NFPA 291 to indicate the water flow available from the hydrant (table 5.1).

When a public water system is not available, developers will need to make arrangements to provide water for their project through other means. They may have

TABLE 5.1
Classification and Marking of Fire Hydrants

Classification	Rated Capacity (gpm)	Rated Capacity (L/min)	Marking Color
Class AA	>1500	>5680	Light blue
Class A	1000–1499	3785–5675	Green
Class B	500–999	1900–3780	Orange
Class C	<500	1900	Red

to install storage systems such as reservoirs or elevated water storage tanks or drill wells to supply water to reservoirs or water storage tanks (figure 5.4). In addition to the water supply system, private water mains and hydrants will need to be installed. Sprinkler systems are required to have at least one valve that allows for the system

FIGURE 5.4 Elevated water storage tanks can be used to store fire protection water for building fire protection demands.

FIGURE 5.5 Each building that has a sprinkler system and is served by a private water system with only one water supply is required to be equipped with a post indicator valve.

to be shut down for maintenance and repairs. Each building served by the private water system that has a sprinkler system and has only one water supply is required to be equipped with an indicator valve (figure 5.5). These valves serve to annunciate whether the water system main valve is in the open or closed position, providing water to the building sprinkler system. There are three types of indicator valves, underground gate valves with indicator post attached; underground butterfly indicating valves with post (both sometimes referred to as the "post indicator valves"); and outside screw and yoke (OS&Y) gate valve. The most common type of indicator valve is the OS&Y valve. The stem on the valve rises and falls depending on whether the valve is open or closed. When the valve shutoff wheel has little or no stem showing above the wheel, the valve is closed (figure 5.6). When the stem is sticking above

FIGURE 5.6 When the valve shutoff wheel has little or no stem showing above the wheel, the valve is closed.

FIGURE 5.7 When the shutoff valve wheel has the stem almost all sticking above the wheel, the valve is open.

the wheel, the valve is open (figure 5.7). OS&Y valves are usually installed on the system riser, which is often found in one of the exit stairwells. Post indicator valves are generally installed outside of structures. A window in the post displays either the word "Open" or "Closed," indicating the position of the valve. Post indicator valves should be monitored or padlocked to prevent tampering. Firefighters preparing to fight a fire in a building with post indicator valves should check the valves to make sure they are in the open position. Other indicating valves need to be supervised or locked as well. Supervision consists of tamper switches hooked into the building fire alarm system.

FIRE HYDRANTS

Early fire hydrants did not have standardized hose connections, which made it impossible for mutual aid departments to hook up to hydrants in other jurisdictions. Several disastrous fires, including the Great Baltimore Fire of 1904, resulted in the establishment of a national standard thread. Firefighters responded from several cities to Baltimore to assist with the fire, which started on Sunday, February 7, 1904. Fire companies arrived from Washington, D.C.; Altoona, Chester, Harrisburg, York, and Philadelphia, Pennsylvania; New York; Annapolis, Maryland; and Wilmington, Delaware, to assist the Baltimore Fire Department. All told there were fifty-seven engine companies, nine truck companies, two hose companies, and one fire boat. Some of the companies responding had hose threads compatible with Baltimore's threads, but most did not. This fact was a major contributing factor to the great damage caused by the fire. Most departments throughout the country today use the national standard thread.

Optimal positioning, spacing, location, and marking of fire hydrants can aid the fire service during emergency operations. Public fire hydrants are often under the purview of a local water authority, many of whom use American Water Works Association (AWWA; http://www.awwa.org/) standards for fire flow and other criteria. The building design team is often responsible for providing hydrants and water supply systems on privately owned property sites. Both the IFC and NFPA 1 include appendices that give criteria for fire flow and fire hydrant location and distribution. Other criteria can be found in NFPA 24, *Standard for the Installation of Private Fire Service Mains and Their Appurtenances.* Private water mains that supply fire hydrants are required to be a minimum of six inches (152.4 mm) in diameter. Private fire hydrants are required to have national standard threads unless approve otherwise by the authority having jurisdiction (AHJ). Typically, hydrants have a large suction hose connection (4½ inches is a common size) called a "pumper outlet" or a "steamer" connection (figure 5.8). Large hydrant connections, sometimes referred to as "steamer ports," are a carryover from the days of horse-drawn steamers. Steamers would hook up to the 4½-inch ports to get water into the steam-operated pumps. In addition, most hydrants normally have two 2½-inch hose connections, although it is not uncommon to still find hydrants in some areas that only have two 2½-inch connections. Where water mains are very small you may encounter hydrants with only a single 2½-inch port. Both wet barrel–type hydrants and the dry barrel types used in areas subject to freezing have these features. Private hydrants are not required to have pumper outlets unless the demand for large hose is added to the attack hose and sprinkler system demands when determining the total demand on the fire protection water supply. Private fire hydrants should not be located less than forty feet (12.2 m) from protected buildings. The number of required private hydrants and spacing is to be determined by the AHJ.

FIGURE 5.8 Typically, hydrants have a large suction hose connection (4½ inches is a common size), called a "pumper outlet" or a "steamer" connection.

Some jurisdictions such as Fallon, Nevada, have taken hydrant connection one step further and have installed Stortz quick-connect hose couplings on the steamer port of the fire hydrants. When responding to a fire, all they have to do is connect the end of the five-inch hose directly to the hydrant. This greatly shortens the time it takes to establish a water supply at a fire scene.

It is not unusual to hear firefighters in rural areas complain that they have difficulty fighting fires because there is no water available. There are actually very few areas of the United States where that statement would be true. Static water supplies are every where; it's just a matter of making them available when a fire breaks out. Sometimes when you are flying on a clear day and happen to have a window seat, take note of all the surface water sources that are available as you fly over almost any part of the country. Dry hydrants are one such method of making static water supplies available for firefighting operations. According to NFPA 1142, *Standard for Water Supplies for Suburban and Rural Firefighting*, dry hydrants are defined as "an arrangement of pipe permanently connected to a water source other than a piped, pressurized water supply system that provides a ready means of water supply for firefighting purposes and utilizes the drafting (suction) capability of fire department pumpers." Dry hydrants often have only a large connection or pumper outlet. Hooking up to a dry hydrant is much quicker than setting up to draft from a static water supply. Usually only one length of hard suction hose is required to hook up to a dry hydrant, and the strainer should already be a part of the dry hydrant system. Dry hydrants should be designed to provide a minimum of 1000 gpm for firefighting purposes. Dry hydrants are permanently installed, utilizing static water supplies such as lakes, ponds, streams, cisterns, and other sources. Any water source used for dry hydrants should be reliable. Dry hydrant locations should have all-weather access available. Dry hydrants have several advantages. They provide an easy-to-build, low-cost solution to locations that do not have a public or private water source for fighting fires. They reduce the distances that water shuttles need to be conducted, saving time and creating more water availability at the fire scene. Like other hydrants, dry hydrants require maintenance. Dry hydrants should be inspected quarterly to make sure the water supply has not deteriorated. Any vegetation that has grown around the hydrant should be removed. Dry hydrants should be painted with a reflective material for high visibility to responding firefighters. Maintenance records should also be kept for inspection by the AHJ.

Optimal location and positioning of hydrants, whether wet or dry, facilitates rapid connection of hose lines and devices. Considerations for designers include height, orientation, distance from the curb, and distance from surrounding obstructions. A clear distance is essential around the hydrant to enable a hydrant wrench to be swung 360 degrees on any operating nut or cap nut. If the nearby obstruction is a plant or bush, consider its potential growth when planning for hydrant placement (figure 5.9). Fire hydrants, like water mains, can be public or private and come in two general types, base valve (dry barrel) and wet barrel. Base valve hydrants are largely used in cold climates because the valve is below ground, below the freezing line so that the hydrant will not freeze. It is important in cold weather to make sure the water drains out of the hydrant barrel before the caps are put back in place to reduce the chances of freezing. The wet barrel type, also known as the California type, can be

FIGURE 5.9 A clear distance is essential around the hydrant to enable a hydrant wrench to be swung 360 degrees on any operating nut or cap nut. If the nearby obstruction is a plant or bush, consider its potential growth when planning for hydrant placement.

used where the temperature remains above freezing most of the year. These hydrants have a valve at each outlet.

Maximum distance between hydrants differs greatly, depending on various local standards. IFC and NFPA 1 both include tables within appendices that enable a designer to find the required fire flow for any given building and then select the corresponding hydrant spacing. IFC bases hydrant distribution on the fire flow requirement for a particular building. Hydrants are spaced accordingly to provide the required fire flow for the building in question. The greater the fire flow required, the less the distance allowed between hydrants. The maximum distance between hydrants and any point on a street or road frontage allowed in the IFC and NFPA 1 is for a 1,750 gpm fire flow or less and allows 250 feet. For a 7,500 gpm fire flow or greater the maximum distance between any point on a street or road, frontage to a fire hydrant is 120 feet. Average spacing between hydrants is five hundred feet where fire flow requirements are 1,750 gpm or less and two hundred feet where fire flow requirements are 7,500 gpm or greater. Where apparatus may approach from different directions, hydrants should be placed primarily at intersections. If additional hydrants are needed to comply with local spacing requirements, they should be spaced along blocks at regular intervals. Fire department pumpers may utilize hydrants in different ways depending on the tactics chosen for a particular fire. If the fire is close enough to a hydrant, a pumper can be positioned at the hydrant and use a large-diameter soft suction hose (soft suction hose is made of the same types of materials as regular fire hose; figure 5.10). Pumpers in urban and suburban areas with hydrants are generally equipped with large-diameter suction hoses connected to an intake on the pumper's front bumper, rear step, or side. This suction hose may be as short as fifteen feet. In many urban areas, however, pumpers carry longer suction hoses in order to reach hydrants on the opposite side of a single line of parallel

FIGURE 5.10 If the fire is close enough to a hydrant, a pumper can be positioned at the hydrant and use a large-diameter soft suction hose to supply water to the standpipe system.

parked cars. If a fire is not close to a particular hydrant, a pumper may have to lay one or more hose lines between the hydrant and the fire. If a pumper lays a supply hose line from a hydrant toward the building with the fire emergency, this is called a "straight" or "forward" hose lay. The opposite (laying supply hose from a building on fire to a hydrant farther down the street) is called a "reverse lay." Many fire departments use one or the other of these tactical options as their standard procedure. Certain conditions, however, may require the use of either type of lay. When I first started out in the fire service as a volunteer in the late 1960s, our hose was loaded so that either a forward or reverse lay could be accomplished easily. Reverse lays were used when additional companies were delayed in response so that the pumper could provide its own water supply to the fire scene quickly. Forward lays were used most of the rest of the time.

Designers should take the types of tactics used by the local fire department into account when locating hydrants. For instance, hydrants placed at the end of dead-end streets will not facilitate straight hose lays very easily, delaying firefighting operations. Hydrants that are too close to a particular building are less likely to be used due to potential fire exposure or collapse. Locating hydrants at least forty feet away from protected buildings is recommended. If this is not possible, consider locations with blank walls, no windows or doors, and where structural collapse is unlikely (such as building corners). A rule of thumb for collapse zone size is twice the distance of the building's height. This is not a consideration in urban areas, where a multitude of hydrants are available for any given location. In private hydrant systems, when hoses are attached directly to them, hydrants are usually spaced so that no hose stretch is greater than 250 feet. The minimum spacing should be such that two hose streams can reach all parts of the building not covered by a standpipe system. When hydrants are installed, care should be exercised to make sure the hydrant is accessible when needed by the fire department. Connections should face the street, and outlets

FIGURE 5.11 Connections on fire hydrants should face the street and outlets should be a minimum of eighteen inches above the ground to allow for working area when connecting hoses.

should be a minimum of eighteen inches above the ground to allow for working area when connecting hoses (figure 5.11). In areas where winter snow can accumulate to a depth that may cover hydrants, flags or other markers can be placed on them to help responding firefighters locate the hydrants. Residents should also be encouraged to shovel snow from around hydrants to make them accessible to the fire department during an emergency. Local building and fire codes, along with NFPA codes, should be consulted when private and public water systems and hydrants are installed.

The National Fire Protection Association has developed a system of classifying and marking fire hydrants so responding firefighters will know generally what the fire flow is of a particular hydrant within an established range. Details can be found in NFPA Standard 291, *Fire Flow Testing and Marking of Hydrants.* Table 5.1 shows the hydrant dome colors indicating the rated capacity or fire flow with a minimum residual pressure of 20 psi. Hydrants are classified according to the fire flow they provide. Class AA hydrants have a rated capacity of 1,500 gpm or greater. Using the NFPA 291 color-coding system, the dome and hydrant caps on these hydrants would be painted light blue. Class A hydrants have a rated capacity of 1,000 to 1,499 gpm Using the NFPA 291 color-coding system, the dome and caps on these hydrants would be painted green. Class B hydrants have a rated capacity of 500 to 999 gpm. Using the NFPA 291 color-coding system, the dome and caps on these hydrants would be painted orange. Class C hydrants have a rated capacity of 500 gpm or less. Using the NFPA 291 color-coding system, the dome and caps on these hydrants would be painted red. Those hydrants with a residual pressure of less than 20 psi should have the pressure stenciled on the dome in black. Hydrants with residual pressures of less than 20 p.s.i. should only be used as a last resort. Hydrant caps should also be painted the same color as the dome. The NFPA recommends that the barrel of the hydrant be painted chrome yellow, but other colors can be used depending on local interests. Hydrants that are permanently inoperative or unusable should be removed.

A number of methods are used to enable firefighters to rapidly identify hydrant locations. The color used for hydrants should contrast as much as possible with the predominating surroundings. Some localities place reflective tape around the hydrant body. Other jurisdictions mount reflectors (usually blue) in the center of the roadway in front of each hydrant; however, in cold weather climates, these reflectors may be obstructed by snow. The best way to identify hydrants in areas subject to snowy weather is a locator pole, which is visible above the highest expected snowfall. These poles can be reflective or contrasting in color, and some have a flag, sign, or reflector mounted on top. These poles should be flexible enough to return to their upright position if someone tampers with them or rigid enough to prevent this type of tampering. Some jurisdictions or sites go so far as mounting a light (usually red or blue) above hydrants. During construction or demolition, fire hydrants may be out of service. Designers should specify that inoperative hydrants be covered or marked during their projects so that firefighters will not waste time attempting to use them.

STATIC WATER SUPPLIES

Two of the most common types of static water supplies are the elevated tank and the ground-level suction tank equipped with fire pumps. NFPA 22, *Standard for Water Tanks for Private Fire Protection*, contains requirements for the design, construction, installation, and maintenance of tanks and auxiliary equipment that provide water for fire protection. Gravity tanks for sprinkler system water supply were very common at one time. Buildings throughout communities had gravity tanks on top of buildings and towers (many still exist, though they may no longer be operative; figure 5.12). Public fire protection water systems across America are fed by water from elevated storage tanks. Capacities of elevated tanks range from 30,000 gallons

FIGURE 5.12 Gravity tanks for sprinkler system water supply were very common at one time. Buildings throughout communities had gravity tanks on top of buildings and towers (many still exist, though they may no longer be operative).

(115 m³) to 500,000 gallons (2000 m³) for steel tanks and 30,000 to 100,000 (380 m³) for wooden tanks. Hydraulic design of sprinkler systems has reduced the usage of elevated tanks in private systems. Elevated tanks have given way to ground-level suction tanks and fire pumps. Static supplies including rivers, ponds, and harbors along with reservoirs may be used to supplement public systems if the supplies are insufficient in volume or pressure or are not dependable. Backyard swimming pools can also be a source of water when no other supply is available.

FIRE DEPARTMENT CONNECTION

Fire department connections (FDC) are installed on buildings with sprinkler systems, standpipe systems, or combination systems to allow the fire department to supplement the buildings normal water supply during fire conditions (figure 5.13). Requirements for fire department connections appear in NFPA 13, *Standard for the Installation of Sprinkler Systems*; 13R, *Standard for the Installation of Sprinkler Systems in Residential Occupancies up to and Including Four Stories in Height*; 14, *Standard for the Installation of Standpipe and Hose Systems*; 15, *Standard for Water Spray Fixed Systems for Fire Protection*; 16, *Standard for the Installation of Foam-Water Sprinkler and Foam-Water Spray Systems*; and 750, *Standard on Water Mist Fire Protection Systems*. These standards set minimum criteria for FDCs, such as which systems require them, their arrangement, and the pipe sizes they feed. The IBC and IFC also contain requirements for FDC location and signage. In some cases, FDCs are not required because they would be of little or no value. Examples include remote buildings that are inaccessible to the fire services, large open-sprinkler deluge systems that exceed the pumping capability of the fire department, and very small buildings. Designers should always seek out and follow fire department requirements, recommendations, and advice for special circumstances. The sole users of FDCs are the fire departments (public and private) that must connect to them. Any deficiency related to the FDC can cause delays in fire suppression and therefore a decrease in the safety of both firefighters and building occupants.

FIGURE 5.13 Fire department connections (FDC) are installed on buildings with sprinkler systems, standpipe systems, or combination systems to allow the fire department to supplement the building's normal water supply during firefighting operations.

FIGURE 5.14 Fire department connections allow the fire department pumper to lay lines from a hydrant to the FDC to supplement water and pressure to sprinkler systems inside protected buildings.

Fire department connections allow the fire department pumper to lay lines from a hydrant to the FDC to supplement water and pressure to wet or dry sprinkler systems, wet or dry standpipes, or both in protected buildings (figure 5.14). No shutoff valves can be installed between the FDC and the fire protection and standpipe systems in the building. Fire department connections are required by code to be within one hundred feet of a fire hydrant. This allows a pumper to hook up directly to a hydrant with its suction hose and then use a preconnected hose line to quickly feed the FDC. For example, if pumpers in a jurisdiction each carry a 150-foot preconnected, 2½-inch hose line, a maximum distance of one hundred feet will enable firefighters to manually stretch this hose to the FDC, regardless of the position of the pumper at the hydrant. If there are multiple FDCs, each should meet this distance requirement from separate hydrants to allow for completely redundant operations. An adequate amount of working room surrounding the FDC will enable a firefighter to approach and connect hose lines without delays. If the inlets to the FDC are a straight type (perpendicular to the wall), a clear path approximately four feet wide would accommodate the firefighter and the hose lines. If the inlets are an angle type, a clear distance of approximately three feet on each side of the FDC will prevent hose lines from kinking (and restricting flow) when they are charged. The designer should consider site conditions leading to the FDC to make it easier for firefighters to stretch hose lines to it without encountering obstructions. Sidewalks, steps, grassy areas, or low ground cover will not slow down this process. However, if a firefighter needs to negotiate walls, climb a ladder, maneuver around a fence or hedgerow, or cut away a bush, the operation will be delayed. Designers should consider the potential growth of nearby bushes or plants and not just the size when planted (figure 5.15). Locations that are likely to be blocked should be avoided. Loading docks, by their nature, are subject to temporary storage and vehicular traffic (figure 5.16). Another example of a poor location would be in front of a supermarket or department store, where stock or carts may block the FDC at any given time. This may

FIGURE 5.15 Designers should consider the potential growth of nearby bushes or plants and not just the size when planted.

be a good reason to deviate from the "street front" requirement or to locate the FDC in a column abutting the road. Designers should always keep in mind how the building will be used, not just how a particular item will look on the construction plans (devoid of people and equipment). Designers should pay special attention to hazardous materials. They should locate FDCs away from fuel tanks, gas meters, or other highly flammable or explosive substances or processes. The designer should also consider the locations of entrances and exits when locating FDCs. A charged hose line is very rigid and will block an outward-swinging exit door or provide a trip hazard for exiting occupants and entering firefighters. Avoid locating FDCs with their inlets pointed in

FIGURE 5.16 Locations that are likely to be blocked should be avoided. Loading docks, by their nature, are subject to temporary storage and vehicular traffic.

FIGURE 5.17 FDCs subject to vehicle damage should be protected by barricades, such as the bollards often used near fire hydrants.

the direction of doors, so that firefighter access and occupant egress is not impeded. A freestanding FDC is an option as long as it is acceptable to the local fire department. Designers may position these anywhere on the property. If an FDC is located far from the building it feeds, consider using special signage. FDCs subject to vehicle damage should be protected by barricades such as the bollards, often used near fire hydrants (figure 5.17). An alternative to protect wall-mounted FDCs is a wall-mounted guard.

The appendix of NFPA 13 recommends that FDCs be installed so that the centerlines of the inlets are between eighteen and forty-eight inches above the adjacent ground (figure 5.18). This height will make hose line connection straightforward.

FIGURE 5.18 The appendix of NFPA 13 recommends that FDCs be installed so that the centerlines of the inlets are between eighteen and forty-eight inches above the adjacent ground.

Some jurisdictions may prefer a maximum height of forty-two or even thirty-six inches. Designers should position FDCs based on the final grade rather than the reverse. If the grade is built up in one area with a mound of soil or mulch to achieve the correct height, this can easily be inadvertently changed later by a landscaper. Or if a platform is built to achieve the correct height, a fall hazard is created for firefighters who may be working in the dark and/or in smoky conditions. Wall-mounted FDCs should be positioned at least forty feet away from windows, doors, or vents. This will minimize the chance that fire, heat, or smoke will make it difficult to connect hose lines. NFPA 13 and NFPA 14 require that a small sign with one-inch raised letters be provided on each FDC to identify the type of system (such as sprinkler, standpipe, or combined). These are frequently cast into the plate surrounding the inlets with raised lettering (figure 5.19). Some jurisdictions require or prefer more prominent marking. Larger signs can be made visible to firefighters and pumper drivers farther away. Icons may be provided to indicate whether the connection feeds sprinklers, standpipes, or both. One example of standard signage for this type of use can be found in NFPA 170, *Standard for Fire Safety Symbols*. Prominent signs can help greatly when the FDC is on a building set back from the street. Some jurisdictions require a light to help identify the FDCs location in the dark.

Pump operators are normally trained to supply a certain amount of water pressure to the FDC to augment the system. For example, standard procedure would be to pump sprinkler systems at 125 psi and standpipe systems at 150 psi. Firefighters may adjust this to provide additional pressure to a higher elevation in a given building or to account for different hose line configurations on standpipe systems. Keep in mind when designing high-rise buildings that at some point the building will be too high for fire department pumpers on the street to boost the water or pressure to the building. Stationary fire pumps would need to be installed in the building to provide the necessary water and pressure. When a sprinkler system requires 150 psi or more, to function properly, NFPA 13 requires that a sign indicate the required pressure.

FIGURE 5.19 NFPA 13 and NFPA 14 require that a small sign with one-inch raised letters be provided on each FDC to identify the type of system (such as sprinkler, standpipe, or combined).

Such a sign alerts the pump operator to this unusual condition. A designer should consider specifying additional FDC signage for underground buildings or transit system facilities. This is because the visual cues that a pump operator typically has on aboveground buildings (such as size or height) are absent. Also, smoke or fire venting provides no indication about the subsurface level where the fire is located. In these cases, a sign indicating the maximum depth and longest horizontal run of pipe gives a pump operator an idea of the pressure he must provide to reach the most remote areas of the system. In some circumstances, an FDC will feed a system covering only a portion of the building (figure 5.20). Signage at the FDC indicating such partial protection alerts responding firefighters to this so they may factor it into their risk analysis and develop appropriate tactics for fire suppression. Signage should provide enough detail so that firefighters connecting the hose lines can identify the proper connection. There are also situations where multiple FDCs on a building are not interconnected. Under these circumstances, designers should consider signage to assist the fire department in supplying the correct FDC. Diagrammatic signs are visually

FIGURE 5.20 In some circumstances, an FDC will feed a system covering only certain buildings or a portion of a building. This is important information for firefighters to know.

FIGURE 5.21 At the University of Maryland, Baltimore, we have a building that sets to the back of a property with four other buildings that has an FDC located on another street away from the building.

the most helpful. FDCs that are far from the buildings they feed also need special signs. If multiple buildings and the FDC locations make it unclear which FDC goes to which building, designers should provide appropriate identification. At the University of Maryland, Baltimore, we have a building that sets to the back of a property with four other buildings. One of the four buildings does not have sprinklers or standpipes (figure 5.21). Yet one FDC sets between the sprinkled and unsprinkled buildings. An additional building is readily visible in the background. It is certainly unclear which building the FDC services. The configuration might also lead firefighters to wonder if the FDC serves all three. As it turns out, the FDC only covers the building on the left. The building on the right is unsprinkled and the one in the background has a sprinkler system, but the connection is on another street (figure 5.22). We have placed signs on all of the connections identifying which buildings they are connected to. We have also placed signs next to the fire department annunciator panels indicating where the FDC for the buildings are located. This information is placed into the fire department information packet for each building as well.

Most code requirements generally call for just one fire department connection. However, NFPA 14 requires multiple FDCs in remote locations on high-rise buildings. This code requirement was added after experience with high-rise fires showed that broken glass and debris falling from a fire area can damage hose lines. There is an exception to the multiple FDCs if it is approved by the local fire department. A second, remote FDC increases the dependability of the water supply. When a building has multiple FDCs, most fire departments would prefer that they be interconnected. This enables the fire department to feed any system from any FDC. However, sometimes this is not possible. For example, a manual dry standpipe system (with no connected water supply) cannot be interconnected with an automatic

FIGURE 5.22 This FDC sets in the courtyard between four buildings. One building does not have a sprinkler system. Two buildings have FDCs on another street. To avoid confusion, the FDC in the courtyard has a sign to indicate which building it serves.

sprinkler system. When dry standpipe systems are installed they are not usually interconnected because of the cost involved. Therefore, each dry standpipe has its own FDC. Dry standpipes are not permitted on high-rise buildings. Sometimes FDC interconnection is not preferable, as previously discussed. When FDCs are not interconnected, the designer should consider special signage as previously discussed, indicating which standpipe the FDC supplies.

Most standards do not specify the number of inlets required on each FDC. NFPA 13 does say that a single inlet is acceptable for FDCs feeding pipe that is three inches or smaller. However, no requirements are identified beyond that. Many FDCs have dual inlets; these are often referred to as "Siamese" connections. One rule of thumb is to provide one inlet for each 250 gpm of system demand, rounded up to the next highest increment of 250 gpm. For example, if the system is 700 gpm, the designer would specify three inlets. Likewise, a system with a demand of 800 gpm would need four inlets. To permit the connection of hose lines, the inlet size and type (threaded or quick connection) must match the type used by the particular fire department. In jurisdictions where the fire service uses threaded hose couplings, FDCs include one or more 2½-inch hose inlets. The thread type will usually be NH Type (American National Fire Hose Connection Screw Thread). To facilitate the connection of the externally threaded (male) end of the fire hose lines, threaded inlets should be the swiveling, internally threaded (female) type. The nonthreaded connections will usually be four or five inches in size. NFPA 1963, the *Standard for Fire Hose Connections*, sets out specific detailed requirements for both threaded and nonthreaded (quick-connect) hose connections. Some buildings are equipped with dry standpipe systems, and the only way they become wet is by supplying the FDC.

FIGURE 5.23 Several different types of caps have been developed to cover the openings to the FDC. One such cap is the male threaded plug that fits into the female opening and can be tightened down. These caps come in brass or chrome finishes to match the finish of the FDC.

FDC inlets are ordinarily required to be provided with threaded plugs or break-away-style caps. Fire department connections are the primary access point for firefighters to get water into a multistoried building to support firefighting efforts. These connections need to be protected against vandalism and people placing objects and trash inside the connection ports, which might obstruct water input. Several different types of caps have been developed to cover the openings to the FDC. One such cap is the male threaded plug that fits into the female opening and can be tightened down. These caps come in brass or chrome finishes depending on the finish of the FDC (figure 5.23). The problem with these caps is that anyone with the strength or a wrench can open them and steal the brass fittings for recycle to obtain money. Aluminum and plastic covers have also been developed to cover the openings (figure 5.24). This type of cover has eye hooks that fit over the nipples on each side of the coupling swivel for the spanner wrench. Once again, though, it does not take much effort to remove the caps. The aluminum is taken for recycle much like the brass by homeless people and other vandals. Plastic has no recycle value, but I theorize that they are removed out of anger because you are not using brass or aluminum that has value when recycled! The Knox® Company, which is known for its Knox® Boxes used for fire department access to building keys, has developed a locking cap for FDCs that operates on the same principle as the Knox® Box. These caps are designed so that they can only be opened with a special key (figure 5.25). Caps are threaded and screw into the female opening on the FDC. Once in place, a special "key" is used to lock down the cap. Only the fire department has access to the keys, and it is difficult if not impossible to remove the caps once in place without the special key.

FIGURE 5.24 Aluminum and plastic covers have also been used to cover the openings on the FDC when the male threaded plug has been stolen or removed by vandals.

FIGURE 5.25 The Knox® Company, which is known for its Knox® Boxes used for fire department access to building keys, has developed a locking cap for FDCs that operate on the same principle as the Knox® Box. These caps are designed so that they can only be opened with a special key.

When building owners use this type of locking FDC cap, they remove the need for code-required periodic inspections of the FDC for caps and obstructions because the caps cannot be removed without the key. Using this system positively secures the FDC and keeps it ready for fire department use when needed. In Baltimore, the University of Maryland became the first location in the city to use the locking Knox® caps. Since their installation over four years ago, there has never been any removal or vandalism associated with the locking caps. I would highly recommend their use.

NFPA standards contain performance language regarding the accessibility of FDCs and the ease with which hose lines can be connected. How the designer meets these requirements can facilitate streamlined fire department operations when a fire occurs. The IBC and IFC specifically require that fire departments approve FDC locations. It is important for designers to seek and obtain this approval. Both NFPA 13 and 14 require that FDCs be on the street side of buildings. The intent is to make them immediately accessible to approaching fire apparatus. Locations on the street side are obvious in urban settings where buildings front directly onto the streets. However, for buildings set back from the street, the street side may be subject to interpretation. In these cases, the designer should consult fire department officials about apparatus approach direction and operational procedures.

STANDPIPES

Firefighters, almost from the first organized fire brigade, have had to bring their water supply to a fire or make use of some source outside of the building to fight a fire. This procedure continued though the construction of one-, two-, three-, and four-story buildings. However, as buildings got taller and taller, it was no longer feasible to reach the fire location from the outside of the building and effectively fight the fire with hand hose lines. Aerial ladders could reach up to the sixth and sometimes seventh floors from the exterior. Hand lines can be hooked to the ladder pipe from a window and water can be supplied to upper floors for overhaul operations and some firefighting. But at some point, interior firefighting can no longer be conducted using hand lines supplied from outside the building. Code requirements brought about as buildings got taller required the installation of standpipes within buildings (figure 5.26). All standpipe installations are required to use only components that are listed by a recognized testing organization such as UL or Factory Mutual. Fire department hose valves attached to standpipes allow firefighters to connect to water supplies from within the building. Standpipe systems consist of a fixed piping system and hose valve connections to preclude the need for long hose lays within tall or large buildings. Water is fed into these systems either through an automatic water supply from the building or manually through a fire department connection. The system delivers water to hose connections throughout the building, which are usually located within enclosed stairwells or exterior stairs. Additional hose connections are located in corridors of large buildings. Fire department information packets mentioned in chapter 1 can identify locations of all hose connections in a building and expedite firefighter access to the connections. Firefighters extend hose lines from these hose connections to conduct interior fire suppression operations. Standpipes are, in effect, a critical component in the supply of water to interior firefighting crews.

Deficiencies associated with fire department hose connections can have disastrous consequences, such as the loss of three firefighters in the 1991 Meridian Plaza fire in Philadelphia. Standpipes are required to be installed inside each required exit stairwell in a building. When more than one standpipe is required, they must be connected together at the bottom. Large buildings may also require additional standpipe

FIGURE 5.26 Code requirements brought about as buildings got taller required the installation of standpipes and fire department hose valves within buildings.

hose connections within the building where the distance between stairwells is too great. When hose valves are installed either in cabinets or in the open in stairwells, care should be taken to ensure that the hose valve is installed at a 45 degree angle to make sure hose does not kink when installed and charged with water. While requirements for standpipe hose valves are generally for high-rise-type buildings and other buildings above four stories, it is also helpful if they are installed in buildings that do not require them, to make access to water easier for the firefighters. Fire prevention inspectors and plans reviewers can ask for hose connections even if not required by code because they are needed by firefighters to fight fires in the buildings. Other circumstances that dictate the use of hose valves is large one- and two-story buildings that have large square footage and excessive distances from the outside access doors to the most remote interior points of buildings. Large warehouses would be a good example. Wal-Mart Corporation has a warehouse in Bentonville, Arkansas, that covers several acres. Without interior hose connections, it would be very difficult to fight a fire deep in the building. Having fire department hose valves in these types of facilities makes it easier for firefighters to reach the seat of the fire using interior fire hose connections.

NFPA 14 is the *Standard for the Installation of Standpipe and Hose Systems* and contains information on the design, installation, maintenance, and testing requirements for these systems. There are several different types of standpipe systems based upon their intended use including automatic dry, automatic wet, semiautomatic-dry, manual dry, and manual wet.

Automatic dry systems are dry standpipes that are normally filled with air pressure. When the hose valve is turned on, air is released, and a dry pipe valve is allowed to open, filling the system with water. The water supply for the system must be able to supply the system demand. FDCs are installed on the system side of the control valve and the supply side of the dry pipe valve. Automatic dry systems would be used in areas where piping is subject to freezing. The downside is the delay in water availability when the hose valve is first turned on.

Automatic wet systems are wet standpipes with water available from a water supply capable of supplying the system demand automatically. FDCs are installed on the system side of the control valve and on the supply side of any isolating valves installed. When the hose valve is turned on, water is immediately available. Therefore, automatic wet systems can only be used in locations where piping is in a heated environment.

Semiautomatic dry systems are equipped with deluge valves, which are activated by a remote control device located next to the hose connection. Upon activation of the remote control device, the system is filled with water from the deluge valve. I have seen manual fire alarm pull stations utilized as the control device for "charging" the piping with water. This practice can be confusing if the pull station is activated by building occupants expecting the fire alarm system to operate, especially when no fire alarm is activated by the pull station. Water supply for the system must be able to meet the system demand. FDCs are installed on the system side of the deluge valve.

Manual dry systems do not have an attached water supply. They rely on the fire department supplying water from a fire hydrant and charging the system through the FDC using a fire department pumper. FDCs are directly connected to the system piping.

Signs should be placed on these standpipes by the FDC to notify firefighters that the system is totally dry and they need to supply water. Manual dry systems are not allowed in high-rise buildings.

Manual wet systems have water in the standpipe riser fed by a limited water supply that does not meet the demand of the system. Like the manual dry systems, water must be supplied into the system through the FDC using a fire department pumper hooked to a fire hydrant. Once again, signs should be used to notify firefighters of the conditions present. FDCs are installed on the system side of the control valve and on the supply side of any isolating valves installed.

Both wet and dry manual systems need to be marked with signs that read "MANUAL STANDPIPE FOR FIRE DEPARTMENT USE ONLY." By code, manual standpipes cannot be used for Class II or Class III standpipe systems. Standpipe risers are allowed by code to be combined with sprinkler system risers. Regardless of the type of system to be installed within a building, codes also require that temporary standpipes be in place during building construction and marked with clear signage to let the fire department know where the temporary standpipe FDC is located. Designers should consider the location and marking of temporary FDCs for temporary standpipes to assist the fire service. When specifying the location of a temporary FDC, the designer should consider using areas around the construction site perimeter. If the FDC is located well away from areas likely to be used for storage, unloading, and heavy equipment such as cranes, it is more likely to be accessible for fire department use. The designer should coordinate the temporary FDC location with any planned construction barricades. Fire service operations will be delayed if walls or fences need to be breached to supply the FDC. Very prominent signs should mark the locations of temporary FDCs. During construction, aesthetics are not of great importance. A large, brightly colored sign with "FDC" painted in a contrasting color will help the fire service locate the FDC rapidly amid the clutter of a construction site. Designers should also specify the removal of the sign when the temporary FDC is removed. The fire department's fire prevention bureau should be consulted to determine where the fire department prefers the temporary FDC be located.

Standpipe systems are classified according to usage and further divided into subclasses according to the projected usage of the system once installed. Classes are identified as I, II, and III. Class I standpipe systems are designed for 2½-inch (65-mm) hose lines and are only for fire department use or use by those trained in heavy hose line operation (figure 5.27). In most cases they are only meant for fire department use. Class II standpipe systems are designed with 1½-inch (40-mm) connections primarily for use by building occupants (figure 5.28). For light hazard occupancies, the code allows for the use of 1-inch (25-mm) hose lines if they are listed for the purpose and approved by the authority having jurisdiction. They will usually have a hose rack with hose and nozzle attached for immediate use. If this type of system is used, occupants should be trained on when to use the hose systems. OSHA may also require that occupants meet the competencies for a fire brigade and have proper protective clothing. When this type of system is proposed during the plans review process, building owners should fully understand the implications of this system once installed. Class II systems with hose require periodic maintenance of the hose lines to make sure they are always ready to use safely. Class III standpipe systems are combinations

FIGURE 5.27 Class I standpipe systems are designed for 2½-inch (65-mm) hose lines and are only for fire department use or use by those trained in heavy hose line operation.

of Class I and II (figure 5.29). Usually there are two valves associated with a Class III system, one 1½ inch (40 mm) for the occupants and one 2½ inch (60 mm) for the fire department and/or local fire brigade. For light hazard occupancies, the code allows for the use of 1-inch (25-mm) hose lines if they are listed for the purpose and approved by the authority having jurisdiction. Class I systems used in buildings not classified as high-rise can be manual, automatic, or semiautomatic. Class I systems used in high-rise buildings are permitted to be automatic or semiautomatic. Manual standpipe systems rely totally on a fire pumper supplying water and pressure to the system at the hose valve. Automatic standpipe systems are connected directly to a water supply at all times and require no further action other than opening the valve to supply water to hose lines. Semiautomatic standpipes require the activation of a control device before water is available at the hose connection.

Class I and III standpipe systems require a minimum flow of 500 gpm (1893 L/min) at the hydraulically most remote standpipe hose connection. Class II systems require a flow rate of 100 gpm (379 L/min) at the hydraulically most remote hose connection. The use of Class II and III systems has declined over the years due to the training and equipment requirements associated with them. The majority of systems installed today are Class I. There are some exceptions in the code to the above information, so plans examiners need to pay attention to the project during design to make sure there is a usable system in place when the construction is finished. Hose connections in Class I systems are typically 2½-inch threaded outlets. They can also be reduced with fittings to 1½ inch, which is the typical coupling size of 1½ and 1¾-inch fire hose used for initial suppression operations. It is essential that hose connection type and size match that used by the fire department in the jurisdiction where the building is located. At the University of Maryland, Baltimore, we removed all hose from Class II and III systems in 1997 and currently all hose valves are Class I.

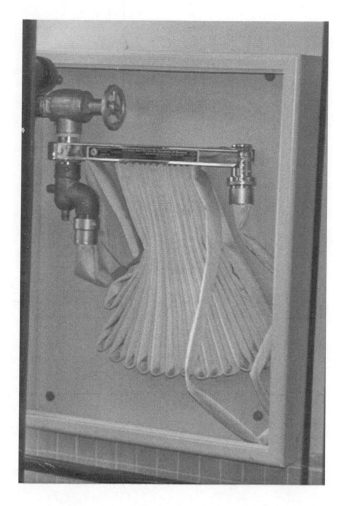

FIGURE 5.28 Class II standpipe systems are designed with 1½-inch (40-mm) connections and single-jacketed fire hose primarily for use by building occupants.

In addition, we added 2½ to 1½-inch adapters on all of the valves so that firefighters can quickly attach attack lines to the valves without using a reducer (figure 5.30).

Building and fire codes specify when designers should incorporate standpipe systems. This can be a locally written code or an adopted model code such as the IBC, the IFC, NFPA 1, NFPA 5000, or NFPA 101. Standpipe systems requirements are based on building height or interior travel distances. In addition, standards such as those issued by OSHA require standpipe systems in certain situations. The IBC and IFC include water supply requirements and some design details. The complete installation standard for standpipe systems is NFPA 14, *Standard for the Installation of Standpipe and Hose Systems*. This standard allows options for hose connections, valves, and other design features. The primary location for hose connections is within enclosed fire-resistance rated stairs. Firefighters set up and begin their attack

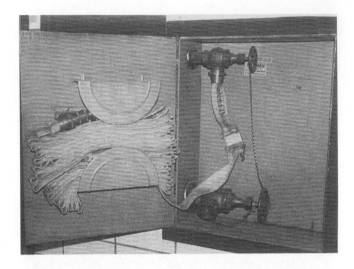

FIGURE 5.29 Class III standpipe systems are combinations of Class I and II. Usually there are two valves associated with a Class III system, one 1½ inches (40 mm) with hose for the occupants to use and one 2½ inches (60 mm) for the fire department and/or local fire brigade.

from within the protected stair enclosure. Then the attack may proceed toward the fire location. If a quick evacuation becomes necessary, the hose then functions as a lifeline, leading the firefighters back to the protection of the stairs. The current suggested location for stairway hose connections in the codes is at the intermediate stair landings between floors (this is subject to local fire department preferences, as all

FIGURE 5.30 These 2½ to 1½-inch adapters were added to all of the fire department hose valves at a major university so that firefighters can quickly attach attack lines to the valves without using their own reducers.

jurisdictions do not prefer this location). Locating fire hose connections at this location is suggested because firefighters usually stretch hose from below the fire floor for their protection. If the connections are at intermediate landings, the hose line reaches farther than it would if the connection were at the main landing, a full story below the fire floor. However, both NFPA 14 and the IBC permit connections to be located at main floor landings if so desired by a given jurisdiction. If hose valves are located on main landings, consider the position of the hose connections in relation to the stairwell entry door. The connections should not be located behind the door when it is open. Designers should position the outlet to permit the hose line to run through the door without kinking and without obstructing egress travel on the stairs. Fire attack using hose lines from stairway hose connections requires stair doors to be propped open. This prevents the hose from becoming kinked and restricting water flow. Open doors can also allow smoke and heat to enter the stairway. However, in high-rise buildings, stairs are pressurized in newer buildings, which will help reduce the amount of smoke and heat that enters that stairwell. At the point when fire suppression operations begin, occupants should either have exited the building, be below the level of the fire, use another stairway, or be sheltered in place until after the incident. However, there is now some concern within the fire protection community that occupants may be exposed to fire or smoke conditions during firefighting operations. Some reasons for this include conflicting evacuation instructions, occupants not following evacuation instructions, delays in evacuation by certain people, the need for the fire department to operate from all stairways, or the need for total building evacuation (especially in response to terrorist incidents).

Remote hose connections outside of stairwells can often be hard to locate (figure 5.31). Hose connections are placed outside of stairwells when distances between stairwells are excessive and normal hose lines will not reach. Locations of remote hose connections should be as uniformly placed as possible on all floors to

FIGURE 5.31 Remote hose connections outside of stairwells can often be hard to locate.

make them easier to find. Highly visible signs or other markings can assist firefighters in locating them quickly. Building information packets can also benefit firefighters trying to locate remote hose connections. Locations of remote hose connections are sometimes tailored to décor or occupancy to satisfy architects or interior designers. This is where the fire prevention bureau needs to step in and make sure the remote connections are properly located for optimum fire suppression efforts. NFPA 170, *Standard for Symbols for Use by the Fire Service*, contains symbols for marking standpipe outlets (hose connections). Placement of remote hose connections can also affect their accessibility. For instance, in parking garages, designers should try to locate hose connections adjacent to drive aisles. Where they are intermingled with parking spaces, an access path at least three feet wide delineated with bollards or a raised, curbed area should be provided to preclude cars from obstructing the connection.

Most new standpipe systems are designed by hydraulic calculations. This ensures that the water supply, pipe sizes used, and pumps (if needed) will provide a certain flow and pressure at a specified number of hose connections in the system. The current NFPA 14 Standard specifies a minimum design pressure for Class I systems of 100 psi at a specific flow rate, which depends on the number of hose connections per floor. However, it includes an exception that allows design pressures as low as 65 psi if this will accommodate the fire suppression tactics. The local fire department needs to be consulted to determine that 65 psi is acceptable to them. Most modern automatic nozzles are designed for optimum operation at 100 psi. Minimum pressures are based on certain assumptions about the fire department equipment and tactics as well as the fixed fire pump feeding the standpipe system. The designer should compensate if the equipment or tactics vary from these assumptions in a particular building or jurisdiction. This will ensure the adequacy of the fire streams to assure the safety of firefighters conducting interior operations. A straight stream nozzle requires at least 50 psi to operate. With the friction loss in fire hose added, 65 psi at the hose connection will provide 50 psi to a straight stream nozzle with 250 gpm flowing through one hundred feet of 2½-inch fire hose. The same pressures can deliver 95 gpm through one hundred feet of 1¾-inch hose. In 1993, NFPA 14 changed the minimum required design pressure from 65 psi to 100 psi at the hose connections. At the same time, this standard was revised to permit longer distances between hose connections and remote areas of a building. Currently, this distance can be up to 150 feet for buildings without complete sprinkler protection and up to two hundred feet for fully sprinkled buildings. The 100 psi design pressure will permit greater flows or longer hose lines but only with the same straight stream nozzles.

Many fire service organizations begin their attacks with fog or combination nozzles that generally require at least 100 psi to operate. This dramatically increases the pressure requirements at the hose connection. If 100 psi is actually available at the connection, every combination of hose size and length will result in inadequate nozzle pressure. It is assumed that firefighters will use fog or combination nozzles early in a fire situation when only one or two hose lines are in operation. It is further assumed that the total flow will be less than the rated flow of the pump. At these lower flows, output pressures will be higher. Finally, it is assumed that if the fire grows, either straight stream nozzles will be utilized or the pumpers supplying the fire department connections will provide greater pressures. Designers must be aware of this information for a number of reasons. First, designers should only use 65 psi minimum design pressure when a particular fire department so specifies, based on their equipment and tactics, not on the cost savings from using a smaller fire pump. An example would be if a department used only 2½-inch hose and straight stream nozzles for standpipe operations. Other design conditions such as additional fire hose connections to enable shorter hose lines may also factor into the decision. In all cases where lower pressure is not specifically approved by the fire department, 100 psi basic pressure should be considered the minimum. However, if any of the above assumptions about the fire pump or the fire department equipment and tactics are invalid in a particular building or jurisdiction, designers should consider providing pressures greater than the basic 100 psi at the hose connections to facilitate adequate fire streams.

Prior to 1991, pressure-regulating devices (PRDs) were virtually unknown to the fire service. The fire at One Meridian Plaza in Philadelphia on February 23, 1991, that resulted in the deaths of three firefighters led to the inclusion of pressure-regulating valves (PRVs) in high-rise standpipe operations training in the fire service. Following the fire, the NFPA issued an alert bulletin (91-3) "Pressure-Regulating Devices in Standpipe Systems" in May 1991. This bulletin warned of the dangers posed to firefighters by improperly set PRVs and provided a much needed wake-up call to the fire service. However, to this day some sixteen years after One Meridian Plaza, many firefighters still do not understand PRVs and the potential implications to firefighting operations. Proper design of PRDs is imperative so that firefighters have adequate pressure for hose streams.

A fire on the twenty-second floor of the thirty-eight-story Meridian Bank Building, also known as One Meridian Plaza, was reported to the Philadelphia Fire Department on February 23, 1991, at approximately 8:40 p.m. and burned for more than nineteen hours (figure 5.32). The fire caused three firefighter fatalities and injuries to twenty-four firefighters. The twelve alarms brought fifty-one engine companies, fifteen ladder companies, eleven specialized units, and over three hundred firefighters to the scene. It was the largest high-rise office building fire in modern American history—completely consuming eight floors of the building—and was controlled only when it reached a floor that was protected by automatic sprinklers. There were several contributing factors that may have resulted in the three firefighter deaths. Partially at fault was inadequate water pressure caused in part by improperly set pressure reducing valves on standpipe hose outlets. The normal attack hose lines used by the Philadelphia Fire Department incorporate 1¾-inch hose lines with automatic fog nozzles designed to provide variable gallonage at 100 psi nozzle pressure. The pressure-reducing valves in the standpipe outlets provided less than 60 psi discharge pressure, which was insufficient to develop effective fire streams or overcome the friction loss in fire hoses. The PRVs were field adjustable using a special tool. However, not until several hours into the fire did a technician knowledgeable in the adjustment technique arrive at the fire scene and adjust the pressure on several of the PRVs in the stairways. When the PRVs were originally installed, the pressure settings were improperly adjusted. Index values marked on the valves did not correspond directly to discharge pressures. To perform adjustments, the factory

FIGURE 5.32 A fire on the twenty-second floor of the thirty-eight-story Meridian Bank Building, also known as One Meridian Plaza, was reported to the Philadelphia Fire Department on February 23, 1991, at approximately 8:40 p.m. and burned for more than nineteen hours.

and field personnel had to refer to tables in printed installation instructions to determine the proper setting for each floor level. The pressure-reducing valves in the vicinity of the fire floor (floors 18 through 20) were set at "80" on the valve index, which corresponded to a discharge pressure between 55 and 57 psi, depending on the elevation. This would provide a nozzle pressure of 40 to 45 psi at the end of a 150- to 200-foot hose line. Several fire department pumpers were connected to the fire department connections to the standpipe system in an attempt to increase the water pressure. The improperly set PRVs effectively prevented the increased pressure in the standpipes from being discharged through the valves. The limited water supply prevented significant progress in fighting the fire and limited interior forces to operating from defensive positions in the stairwells.

NFPA 14 imposes a maximum pressure limit of 175 psi on standpipe systems for fire department use. Pressures in excess of 175 psi will invoke requirements for pressure-reducing devices. Pressure-regulating devices restrict system pressures, usually below 175 psi for Class I systems. This is considered the maximum safe operating pressure as well as the maximum working pressure limit of most fire protection components. PRVs are usually located in high-rise buildings in two locations: at floor/zone control valves for combination standpipe/sprinkler systems and at standpipe hose valve outlets. PRDs fall into three categories: pressure reducing valves, pressure control valves, and pressure-restricting devices. Pressure-restricting devices do not limit pressure during static (nonflowing) conditions nor do they maintain a constant discharge pressure. These devices incorporate orifice plates, mechanical pressure restrictors, or valve-limiting stops. Pressure-restricting devices are not used for new Class I standpipe systems. However, designers may encounter these when redesigning existing systems, which would provide the opportunity to implement some or all of the considerations below. PRVs and pressure control valves limit both static and residual (flowing) pressures. However, many of these valves are factory preset to attain specific outlet pressures with specific inlet pressures. It is important for designers to specify the inlet pressure range for valves as well as the desired outlet pressure so that they may be designed properly and then installed on the correct floors. Careful attention during design, installation, acceptance testing, and maintenance ensures that systems with PRDs will function properly.

PRVs and pressure control valves have other disadvantages. Their failure rate has been high, resulting in the addition of testing requirements to NFPA 14. Secondly, many cannot be adjusted by firefighters in the field during the fire or they require special tools and knowledge. Finally, hose connections with these devices cannot be used as backup fire department connections, since water can only flow through a PFD in one direction (discharge). The most reliable means of limiting pressures in stand-pipe systems is to design them to preclude the need for pressure-regulating devices. In shorter buildings, careful attention to the design of pumps and the maximum pressure supplied by incoming water mains can accomplish this. In taller buildings, the same concept can be applied to each separate vertical standpipe zone. Pressure fluctuations in the water supply as well as the full range of fire pump capacity are essential considerations in any building. If the use of PRDs cannot be avoided, certain design features will balance their disadvantages. The easier the valves are to adjust in the field, the faster the fire services can overcome any unforeseen situation. Designers should select valves that can be easily adjusted and specify that identification signs and adjustment instructions be posted at each valve. The tools required to perform field adjustments should be kept in a secure yet accessible location such as the fire command center or a locked cabinet near the fire alarm annunciator. Finally, a supplemental system inlet should be provided at the level of fire department entry. This can be simply an extra hose connection without a PRD on a riser. NFPA 14 recommends a supplemental inlet, especially important for systems with a single fire department connection.

The vertical pipes that feed hose connections are called "standpipes" or "risers." If there are multiple risers, NFPA 14 requires interconnections with supply piping to form a single system, with valves at the point where each riser is fed by the main bulk piping coming from the water supply point. Designers should also put valves on the feed lines to remote or supplemental hose connections. These valves are all called "standpipe isolation valves." The ones on vertical risers are called "riser isolation valves." They allow the fire department to shut off or isolate any given riser or feed that breaks or otherwise fails. Firefighters may then use the remaining standpipes. NFPA 14 requires that riser isolation valves separately control the feed to each standpipe. Sequential valves are not acceptable where a single valve in the bulk main can shut off more than one downstream riser. For risers in stairways, the riser isolation valves should be within the fire-rated stair enclosure to protect firefighters who may need to operate them. Previous editions of NFPA 14 required designers to place the riser isolation valves at the bottom of the risers to make them quickly accessible to firefighters. Fire departments may still prefer that these valves be located on the level that they use for their primary entry. If the bulk feed main is located on a different level, it could be piped up or down to the fire department entry level, where the isolation valve would be placed for that particular riser.

Standpipes should be installed as the construction of a building progresses. These standpipes can be temporary or permanent. Both the IFC and NFPA 241, *Standard for Safeguarding Construction, Alteration, and Demolition Operations*, contain requirements for standpipes during construction. Design documents should indicate the applicable requirements. A marked, accessible fire department connection can suffice as a water supply until building construction progresses to the point at which the water supply system and fire department pumpers can no longer provide adequate pressure to the system. At this point, a temporary or permanent fire pump also becomes necessary. In climates subject to freezing temperatures, it is vital that standpipes in unheated areas be dry systems. Heat tracing and insulation are ineffective protection for dry fire protection systems because water is not normally flowing through the piping. Large dry systems deserve special considerations. As the size of a dry system increases, the time required to deliver water to the remote hose connection increases. This is due to the increased pipe volume that must be filled. The problem can be mitigated by subdividing the system into smaller independent systems, or zones. A disadvantage is that fire department inlet connections to dry systems cannot be interconnected.

6 Portable Fire Extinguishers

INTRODUCTION TO PORTABLE FIRE EXTINGUISHERS

Portable fire extinguishers are provided in buildings as a "first aid" measure in the event that fire breaks out (figure 6.1). They are designed for use by building occupants; however, portable fire extinguishers may also be used by the fire department. Before the type of fire extinguisher is chosen for a building or portion of a building, it is important to know what types of fuels or hazards are present. Fire extinguishers should be chosen based upon the fuel or fuels present in various portions of the facility. Buildings are often classified as to the types or levels of hazards in terms of fuel load. Light or low hazards exist when there are limited combustible materials in the building. These conditions are usually present in offices, churches, schoolrooms, assembly halls, and other similar occupancies. Ordinary or moderate hazards exist when the combustibles present are significant but consist of materials of ordinary burning characteristics or small quantities of combustible that would facilitate rapid fire growth. Ordinary types of occupancy would likely include mercantile storage and display areas, auto showrooms, and parking garages. Extra or high hazards exist where there are substantial quantities of combustibles that could support rapid fire growth and result in a large fire. Such occupancies might include woodworking areas, aircraft servicing areas, and warehouses with high stacks of combustibles. Types of fires and fuels will be discussed in greater detail later in this chapter.

There is a great deal of controversy concerning the installation and use of portable fire extinguishers by building occupants in some parts of the country. Both the National Fire Protection Association (NFPA) and the Occupational Safety and Health Administration (OSHA) have codes and regulations governing the occupancy requirements and installation of portable fire extinguishers. The *International Building Code* refers to the *International Fire Code*, which requires extinguishers for specific hazardous operations such as asphalt kettles, buildings under construction, dry cleaning plants, lumberyards, and others. NFPA 101, the *Life Safety Code*, requires fire extinguishers in certain occupancies, such as apartment buildings, health care occupancies, hotels, business occupancies (including colleges and universities), mercantile occupancies, and others. The *Life Safety Code* does not require fire extinguishers in educational occupancies, day care centers, or assembly occupancies and some other selected locations. OSHA 1910.157 is the regulatory section covering portable fire extinguishers within the OSHA regulations. OSHA requirements apply to the placement, use, maintenance, and testing of portable fire extinguishers used by employees (figure 6.2). Where the employer has established and implemented a written fire safety policy, which requires the immediate and total evacuation of employees from the workplace upon the sounding of a fire alarm signal, OSHA allows for the removal or exclusion of portable fire extinguishers from a building if occupants are not expected to use them.

FIGURE 6.1 Portable fire extinguishers are provided in buildings as a "first aid" measure in the event that fire breaks out.

FIGURE 6.2 OSHA requirements apply to the placement, use, maintenance, and testing of portable fire extinguishers used by employees.

However, if they are provided with the expectation that occupants will use them, then OSHA requires that occupants be trained to use them. NFPA 101 also has requirements for occupant training under certain sections of the code for specific occupancies. While OSHA, the *Life Safety Code*, and fire prevention codes prescribe when fire extinguishers are installed, specific information about installation, operation, and maintenance are found in NFPA 10, the *Standard for Portable Fire Extinguishers*.

NFPA does not take the same approach to portable fire extinguisher requirements as OSHA. NFPA codes dictate when fire extinguishers are required and the type and size of fire extinguishers as well as the location and occupancy where they are required. Requirements for installation are found in NFPA 101, the *Life Safety Code*, and other sections of NFPA standards. NFPA standards also specify when and who needs to be trained to use portable fire extinguishers. It also exempts certain occupancies from portable fire extinguishers when a building is fully sprinkled. Portable fire extinguishers come in many sizes and with options as to the types of extinguishing agents. Building hazards should be thoroughly evaluated before selecting the proper size and type of fire extinguisher and agent. Even if fire extinguishers are not placed in buildings with the intent of occupant use, they can still be used by responding firefighters. When firefighters respond to alarm bells sounding, they do not always bring extinguishing equipment with them into the building to investigate the cause of the alarm. If they discover a small fire during their investigation, they can use a portable fire extinguisher already in the building to extinguish the fire. Portable fire extinguishers can also be used to complete extinguishment of a fire being controlled by an automatic sprinkler system.

When building occupants or firefighters select a portable fire extinguisher to put out a fire, they should be aware of the proper types and sizes of extinguishers that should be used on the five major classes of fire (table 6.1). To firefighters this might seem like a silly statement, but an incident occurred at a major Midwestern university where city firefighters used 32 ABC dry chemical fire extinguishers to extinguish a flammable metal fire (Class D fire)! Either they ran out of extinguishers, the fire ran out of fuel, or somebody found the correct extinguisher for flammable metal fires (Class D; figure 6.3). Responding firefighters should have a working knowledge of portable fire extinguishers and what types to use on certain fuels. Lay people who use portable fire extinguishers need to be trained so they will know when it is safe to use a portable fire extinguisher, what type to use, how to operate the extinguisher, and when it is time to evacuate and leave the fire extinguishing to the sprinkler system or responding firefighters. They also need to be aware of the need to call the

---

TABLE 6.1

Five Classes of Fire

Class A	Ordinary combustibles	Paper, wood, cardboard
Class B	Flammable liquids	Gasoline, diesel fuel
Class C	Energized electrical equip	Electric motors, transformers
Class D	Combustible metals	Magnesium, potassium, sodium
Class K	Cooking oils	Deep-fat fryers, grills

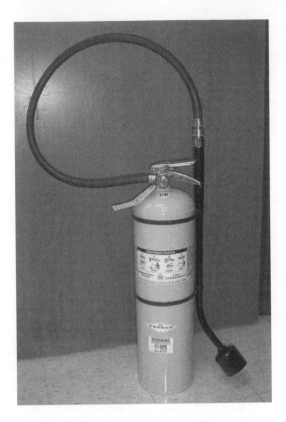

FIGURE 6.3 Class D fire extinguisher designed exclusively for flammable metal fires.

fire department or activate a building's fire alarm system before attempting to use a portable fire extinguisher. Using a fire extinguisher first has resulted in many fires that got out of control before the fire department was called, making their job more difficult than it needed to be.

CHEMISTRY OF FIRE

In order to effectively extinguish fire using portable fire extinguishers, it is helpful to understand what fire is and its physical characteristics. An early attempt to explain the chemistry of fire and to formulate theories for extinguishment was the development of the fire triangle (figure 6.4). The theory behind the development of the fire triangle was that three components are necessary to be present for fire to occur in the first place: oxygen, heat, and fuel. If all

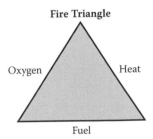

FIGURE 6.4 An early attempt to explain the chemistry of fire and to formulate theories for extinguishment was the development of the fire triangle.

FIGURE 6.5 Oxygen can be excluded from the fuel by "covering" the fuel with dry chemical or foam.

three are present and a fire breaks out, then fire extinguishment occurs when anyone of the three characteristics are removed from the fire equation. Portable fire extinguishers function by removing one or more of the elements of the fire triangle. Oxygen can be excluded from the fuel by "covering" the fuel with dry chemical or foam (figure 6.5). If oxygen cannot reach the fuel source, fire cannot occur. Oxygen can also be excluded from the fire by replacing it with some other gas that will not support combustion (figure 6.6).

FIGURE 6.6 Carbon dioxide extinguishers displace oxygen, thus preventing it from concentrating enough to support combustion, and the fire goes out.

FIGURE 6.7 Water has a tremendous ability to absorb heat but only works effectively on Class A fires.

Carbon dioxide displaces oxygen, thus preventing it from concentrating enough to support combustion and the fire goes out. Heat can also be removed from the fire triangle, which results in the fire being extinguished. Water has a tremendous ability to absorb heat (figure 6.7). Water can absorb 4.184 Joules (J) of energy per degree Celsius per gram of water. That is, if one had one gram of water and raised the temperature of that water by one degree Celsius, then it is said that the water will have absorbed 4.184 J of energy. Water acts by cooling a fire below the ignition temperature of the fuel, thus removing heat from the fire equation, and the fire goes out. Fuel can also be removed from a fire, but only in very limited circumstances. Portable fire extinguishers have nothing to do with the removal of the fuel. If a fire involves a leaking flammable liquid or gas and a shutoff valve is available, then the valve can be shut off, which will remove the fuel from the fire and the fire will go out. Any other attempts to remove fuel would expose users of portable fire extinguishers to unnecessary danger and should not be attempted.

More recently, the theory of the existence of a fire tetrahedron has replaced the fire triangle as the accepted explanation of the chemistry of fire. The fire tetrahedron incorporates all three characteristics of the fire triangle—oxygen, heat, and fuel—and adds the concept of a chemical chain reaction that takes place when fire occurs (figure 6.8). Once again, extinguishment is accomplished with the fire tetrahedron just as with the

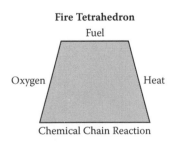

FIGURE 6.8 The fire tetrahedron incorporates all three characteristics of the fire triangle—oxygen, heat, and fuel—and adds the concept of a chemical chain reaction that takes place when fire occurs.

FIGURE 6.9 Multipurpose ABC dry chemical fire extinguishers operate by interrupting the chemical chain reaction that occurs when a fire breaks out and all of the components of the fire tetrahedron are present.

fire triangle: removing one of the four component parts of the fire tetrahedron will cause the fire to go out. Portable fire extinguishers are designed and function by eliminating one of the four components of the fire tetrahedron. We have previously discussed the removal of oxygen, heat, and fuel from a fire, so that will not be discussed further here. The concepts of oxygen, heat, and fuel are the same whether talking about the fire triangle or the fire tetrahedron. Interrupting the chemical chain reaction is the major difference between the theories of the fire triangle and the fire tetrahedron. Interrupting the chemical chain reaction can be accomplished by using a portable fire extinguisher. When a fire occurs because all the elements of the fire tetrahedron are present, multipurpose ABC dry chemical extinguishers can be used to extinguish the fire by interrupting the chemical chain reaction.

Understanding the chemistry of fire will assist those doing fire suppression with portable fire extinguishers in properly selecting and using those extinguishers when a fire occurs. It should be noted that removing the fuel from a fire is the least likely and potentially most dangerous of the options. If a fire involves a piping or tank system, then a valve could be shut off that would remove the fuel from the fire. Flammable liquid and gas tankers have emergency shutoff valves that could be operated to stop the flow of fuel, thus eliminating the fuel for the fire. Drivers of home delivery trucks for flammable propane gas may also carry a remote shutoff control. Fixed storage tanks and containers usually have shutoff valves at the tanks or in the piping that can stop the flow of fuel. Removing the fuel from the fire generally involves shutting off of the fuel from its source. Portable fire extinguishers do not generally operate by removing fuel from fires.

FIRE EXTINGUISHER CLASSIFICATION

Fire extinguishers are classified and rated by the type and amount of fire they are designed to extinguish. There are five major classifications of fires and five types of extinguisher classifications (table 6.1). Class A fires and extinguishers deal with ordinary combustible materials such as wood, paper, cardboard, and plastics.

Class A fire-extinguishing agents can be pressurized water, hand water pump devices, or ABC all-purpose dry chemical fire extinguishers. If either of the water extinguishers is used, it would extinguish fire by cooling the fuel below its ignition temperature. ABC dry chemical extinguishers interrupt the chemical chain reaction of fire, and the fire goes out. When Class A fire extinguishers are installed within a structure or protection area they should be placed no greater than seventy-five feet (22.7 m) travel distance from any point in the structure or area for Class A hazards. Generally, portable fire extinguishers are placed by exterior exit doors or along the natural path of travel to the exits, depending on the size of the structure or coverage area. High hazard occupancies may require more extinguishers be installed at shorter intervals to meet the square footage requirements. Any areas less than the required coverage distances must have a minimum of one portable fire extinguisher, which would be properly placed near one of the exit doors from the protected area.

Class B fires and extinguishers involve fires in flammable liquids and gases, such as propane, gasoline, and fuel oil. Portable fire extinguishers with Class B ratings include ABC multipurpose dry chemical, carbon dioxide, halon, clean agent halon replacements, Purple K®, and foam. ABC multipurpose extinguishers function by interrupting the chemical chain reaction, and the fire goes out. Carbon dioxide fire extinguishers function by excluding oxygen from the fuel. Halon and clean agent replacements function by making the air in the room inert to combustion. Purple K® extinguishing agents function by interrupting the chemical chain reaction. Purple K® extinguishing agents should never be mixed with phosphate-based agents because a chemical reaction takes place that may damage the extinguisher. Foam extinguishing agents blanket the surface of the liquid, excluding oxygen from the fuel. Travel distances to reach fire extinguishers where Class B hazards exist should not exceed fifty feet (15.25 m) from any location in the structure or area.

Class C fires and extinguishers are those that occur and are used with energized electrical equipment. Before fighting energized electrical equipment fires, however, the power should be turned off if at all possible. It is important to note that once electrical equipment is deenergized, the fire becomes a Class A fuel type of fire. Class C fire extinguishers usually involve agents that will not damage expensive electrical equipment when discharged at a fire. They include carbon dioxide, halon, and clean agent halon replacements. Carbon dioxide excludes oxygen from the fire, extinguishing the fire. Halon and replacements make the area where the fire is burning inert to combustion, extinguishing the fire. Multipurpose ABC dry chemical extinguishers will also extinguish Class C fires; however, keep in mind that they will create a huge mess and possibly damage sensitive electrical equipment. ABC dry chemical extinguishers function by interrupting the chemical chain reaction. Water and foam extinguishers should never be used to fight Class C fires involving energized electrical equipment. Water conducts electricity and could expose the extinguisher user to electrical shock, which could result in serious burns or death. There are no fire extinguishers rated for just Class C fires. Extinguishers that also have a Class C rating are placed within structures or coverage areas according to either the Class A or Class B requirements based upon the hazards present.

Class D fires are those involving flammable metals. Flammable metals including magnesium, lithium, sodium, potassium, and others are water reactive. The physical form of the metals is usually small, including chips, shavings, and powders. The smaller the physical form of the flammable metal fuel, the greater the flammability hazard of the materials. Powdered metals can also present an explosion hazard. Extinguishers and agents for Class ABC fires mentioned previously are not effective against Class D flammable metal fires. Water can cause violent explosions when used to extinguish some flammable metal fires. And while water may be used through large-volume hose streams to control or extinguish fires involving flammable metals, portable extinguishers using water or foam should never be used on Class D fires. Class D fires require a unique application technique. Unlike Class A or Class B fires, you will not see a lot of flame or feel a lot of radiant heat in the early stages of the fire. There will be intense light in the case of magnesium, volumes of dense smoke from lithium, and very little smoke with either titanium or zirconium. But do not let the small size of the fire or the lack of flame fool you. These are serious fires and the potential for disaster exists if you underestimate them. Burning Class D material generates hydrogen gas when it is exposed to water, causing a violent explosion. Moisture in the ground, on concrete, or even in some agents themselves may cause this reaction. Therefore, extreme caution must be used when fighting these fires. The control of flammable metal fires is achieved by attacking two fronts simultaneously: excluding oxygen from the combustible metal by completely surrounding it, even what is not yet burning, and by the absorption of heat to below the temperature required to sustain the fire. This latter feature of Class D agents is frequently referred to as "heat sink." The principle of extinguishment is to completely cover the burning material with a layer of Class D agent up to two inches thick. Extinguishers for Class D hazards should be placed no farther than seventy-five feet (22.7m) travel distance from any point in the building or area.

Class K is a relatively new classification system for fires involving cooking where combustible fuels such as animal and vegetable oils are present. These types of materials were previously classified as Class B fuels. In the past, sodium bicarbonate and potassium bicarbonate dry chemicals were specified for use on cooking media. All cooking media, whether animal or vegetable, liquids or solids, contain saturated fats or free fatty acids. When an alkaline-based extinguishing agent is applied to heated saturated fats, a reaction occurs called "saponification." The reaction forms soapy foam on top of the surface that smothers the fire and contains the flammable vapors and the hot cooking oils. Both the dry chemical agents and the wet chemical agents (alkaline based) will cause the same reaction. The wet chemical, applied as a fine mist, has the added advantage of cooling the cooking medium and lowering the temperature, making the wet chemical agents more effective than the dry ones. ABC dry chemical (monoammonium phosphate) is acidic in nature and therefore will not saponify when applied to burning cooking media. It may even be counterproductive when used after alkaline agents have been used, disturbing or removing the soapy foam from the surface, which could cause reignition. Considered a specialty agent, wet chemical (potassium acetate) is highly effective in extinguishing Class K fires, particularly those found in deep-fat fryers. These fire extinguishers are typically required in a commercial kitchen setting. Class K extinguishers are also equipped with specially designed nozzles that will not spread grease fires when discharged. When Class K extinguishers are placed, the maximum travel distance should not exceed thirty feet (9.15 m) from the hazard to the extinguisher. The 2002 edition of

NFPA 10 requires that "existing dry chemical extinguishers without a Class K rating that were installed for the protection of Class K hazards shall be replaced with an extinguisher having a Class K listing when the dry chemical extinguishers become due for either a 6 year maintenance or hydrostatic test."

FIRE EXTINGUISHER RATINGS

Fire extinguishers are generally selected and placed depending upon the hazards of the area protected. Hazards are based on the classes of fire previously discussed. Fire extinguisher ratings are based on the hazards and the amount of agent required to extinguish a certain volume of fire in an area based upon the hazard. Fire extinguishers are given ratings to determine their ability to extinguish a given size of fire. Numerical ratings given to Class A, B, and C portable fire extinguishers indicate how large a fire an experienced person can put out with the extinguisher. The higher the numerical value of the rating, the more fire that can be put out if the extinguisher is used effectively. Ratings are based on tests conducted at Underwriters Laboratories. For example, a 2-A rated fire extinguisher will have approximately two times more fire extinguishing capability than a 1-A rated fire extinguisher. The same is true for the BC rated extinguishers. Typical ratings for portable fire extinguishers are shown below:

- Class A: 1-A, 2-A, up to 40-A
- Class BC: 1-B, 2-B, up to 640-B

Oftentimes you hear people referring to portable fire extinguisher size by weight rather than rating. The rating gives a much more accurate indication of the ability of the fire extinguisher than the weight. Typical sizes of portable fire extinguishers used for given occupancy hazards are shown below:

- Light hazard   4-A
- Ordinary hazard   10-A
- Extra hazard   20-A
- Low hazard   5- to 10-BC
- Moderate hazard   10- to 20-BC
- High hazard   40- to 80-BC

Ratings will be found on the fire extinguisher label. The minimum rating permitted by the fire codes for Class A type fires is 2-A. They do permit using two 1-A water extinguishers to meet the requirement. However, if you are going to go that route, you might as well get an all-purpose ABC dry chemical because water extinguishers alone provide a very limited fire extinguishing capability. They are only rated for Class A fires.

TYPES OF FIRE EXTINGUISHERS

Multipurpose ABC dry chemical fire extinguishers are by far the most common type of fire extinguisher in use today (figure 6.10). A dry chemical agent is placed in the extinguisher container and the container is pressurized with a gas, usually nitrogen,

FIGURE 6.10 Multipurpose ABC dry chemical fire extinguishers are by far the most common type of fire extinguisher in use today.

which is inert, and the nitrogen expels the dry chemical out of the extinguisher. Nitrogen or other propellant can be placed in the same chamber as the dry chemical, although some extinguishers have separate gas and agent containers; the gas container has to be punctured with a special lever installed on the extinguisher body to "charge" the extinguisher before use (figure 6.11). This operation will slightly delay the response time in attempts to extinguish a fire. Monoammonium phosphate is the chemical agent used in most multipurpose ABC fire extinguishers and is generally nontoxic, nonreactive, and nonflammable. Persons with preexisting respiratory problems may be affected by the chemical powder when it is aerosolized during discharge. If dry chemicals are discharged in a confined area, they can reduce visibility and cause disorientation. Dry chemical agents are not conductive, and residue left on electrical contacts may reduce or eliminate the ability of the contacts to conduct. Dry chemical agents can also clog air or air-conditioning filters as well. As the name multipurpose ABC states, the extinguishers are rated for A, B, or C type fires. They should not, however, be used on Class D or Class K fires because they would be ineffective. Other types of dry chemicals in use include sodium bicarbonate and potassium bicarbonate (Purple K®). Sodium bicarbonate is commonly used for Class K fire extinguishers and Purple K® is used for flammable liquid fires. Purple K® is an extremely effective fire-extinguishing agent and acts very quickly. Because Purple K® is expensive (two to three times more than ABC dry chemical), it is generally used by crash rescue at airports and for other specialized protection applications where the cost of the equipment being protected outweighs the cost of the extinguishing agent. Because these

FIGURE 6.11 Some types of ABC dry chemical extinguishers have separate gas and agent containers; the gas container has to be punctured with a special lever installed on the extinguisher body to "charge" the extinguisher before use.

agents are discharged under pressure and rated for flammable liquid fires, care should be taken to prevent the spread of the flammable liquid caused by the pressure from the extinguisher. Dry chemical agents extinguish fire by interrupting the chemical chain reaction and coating flammable materials and container surfaces to prevent reignition. Dry chemical fire-extinguishing agents are used for the following types of fires: multipurpose is used on ordinary combustibles, flammable or combustible liquids, flammable or combustible gases, combustible solids, electrical hazards, and textile operations subject to flash surface fires. Dry chemical agents are generally not effective against chemicals containing their own oxygen; combustible metals such as sodium, potassium, magnesium, titanium, and zirconium; or deep-seated fires in ordinary combustibles where the dry chemical cannot reach the seat of the fire. In addition, multipurpose dry chemical should not be used on machinery such as carding equipment in textile operations and delicate electrical equipment, computers, electronics, and others. Bicarbonate of soda is generally more effective with fires involving flammable animal or vegetable oils in cooking operations.

FIGURE 6.12 Water-based extinguishing agents include water, antifreeze, loaded stream, wetting agent, wet chemical, and foam.

Water-based extinguishing agents include water, antifreeze, loaded stream, wetting agent, wet chemical, and foam (figure 6.12). All but foams are used for Class A fuels only. Water-based extinguishers originally involved three basic designs: stored pressure, pump tank, and inverting. Manufacture of inverted extinguishers was discontinued after 1969. Currently there are two types of water-based extinguishers remaining, stored pressure and hand pump tank. Stored pressure water extinguishers are simply water in an extinguisher (container) with compressed air used to expel the water from the container. Water extinguishers have been around for quite some time. In the beginning, water was gathered in leather buckets and thrown on a fire, sometimes by deploying a bucket brigade (figure 6.13). Hand pumps were developed and are still in use today in some public locations and by fire departments for wildland firefighting. There are two basic types of hand pumps, cylindrical and backpack. Backpacks can be made of metal or plastic (figure 6.14). Backpack types of hand pumps are used exclusively for fighting outdoor types of fires such as brush and other wildland fires. They can contain up to five gallons of water. Cylindrical pump cans may still be found in use in certain locations (figure 6.15). Truck companies often carry pump cans (sometimes referred to as just the "can"). Pump cans typically contain 2½ gallons of water. The water is expelled from the can through the use of a hand pump through a hose and nozzle attached to the can. Water-based extinguishers other than foam are only rated for Class A type fires. They should not be used on Class B flammable liquid fires because they may not be effective or may spread the fire. Water should also not be used on Class C electrical fires because water conducts

FIGURE 6.13 In the beginning, water was gathered in leather buckets and thrown on a fire, through the efforts of a bucket brigade.

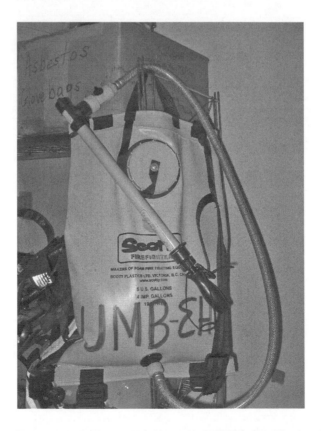

FIGURE 6.14 There are two basic types of pump cans, cylindrical and backpack. Backpacks can be made of metal or plastic.

FIGURE 6.15 Cylindrical pump cans may still be found in use in certain locations.

electricity and presents a potential shock hazard. Many of the metals in Class D fires are water reactive, so water extinguishers should not used on metal fires either. Class K fires involve flammable grease fires. Water is not effective on burning grease and may spread the fire. While water extinguishers are still listed and approved for Class A type fires, they are really becoming obsolete in many locations because of the more practical ABC dry chemical. If water extinguishers are used, there also need to be additional extinguishers to handle Class B and C fires.

Portable foam fire extinguishers are a specialty type of extinguisher and are not commonplace, so they will not be encountered frequently. They resemble pressurized water extinguishers in appearance. Foam is water based, so using foam on energized electrical equipment is not recommended. Two basic types of foam are used in portable fire extinguishers: aqueous film-forming foam (AFFF) and film-forming fluoroprotein foam (FFFP). Extinguishers are available in 2½ gallon sizes and are pressurized to expel the foam agent. The 2½ gallon size is rated at 3-A:20-B and 33 gallon units are available that are rated at 20-A:160-B. When foam is used on Class A fires, it acts to cool and penetrate to reduce temperatures of the fuel below the ignition temperature. When used on Class B fires, they coat the surface of the fuel, therefore excluding oxygen from the fuel, which will extinguish the fire. Water and foam extinguishers can be affected by freezing temperatures. Therefore, the ambient conditions of the space where the extinguisher is placed should be considered. ASME standards require placement of water-based fire extinguishers in ambient conditions between 40 and 120°F (4 and 49°C).

Carbon dioxide (CO_2) is a liquefied gas at normal temperatures and pressures and is used as an extinguishing agent in some portable fire extinguishers with a Class BC rating. Because it is a gas, carbon dioxide does not leave a lasting residue when discharged like the dry chemical agents do. When discharged, CO_2 leaves a snow-like material on surfaces, which is frozen carbon dioxide gas (dry ice). This material quickly evaporates as it warms up and leaves no lasting residue and does not usually harm surfaces. Carbon dioxide is nontoxic, nonreactive, and nonflammable; however, it can displace oxygen in the air and cause asphyxiation at concentrations required for fire extinguishment. Carbon dioxide fire extinguishers should not be used in confined areas without self-contained breathing apparatus (SCBA) for protection. Portable fire extinguishers utilizing carbon dioxide are recognized by the characteristic cone or horn used as a discharge nozzle. As carbon dioxide is discharged from the extinguisher, the gas is very cold. The horn/nozzle and associated hoses or piping will "frost" up during the discharge and can be a cold hazard if touched by bare skin. Carbon dioxide fire extinguishers are generally used in mechanical and electrical spaces, computer rooms, and other areas where BC hazards exist and a clean agent is required to limit cleanup and potential damage. Some older carbon dioxide extinguishers may have metal horns. Since carbon dioxide is rated for energized electrical equipment, this metal horn could cause an electrical shock. Metal horns should be replaced with nonmetallic plastic horns. Discharge of a gas from a container can cause a static electricity buildup. Contact with the horn can cause a shock. This shock may startle the operator but causes no physical harm.

Dry powder is a specialized extinguishing agent used for Class D combustible metal fires and should not be confused with dry chemical. Dry powder functions by forming a crust and smothering the fire. Dry powder can be placed in fire extinguishers or may be encountered in boxes or buckets for manual application to Class D fires.

HALOGENATED AGENTS

Although halogenated fire extinguishing agents are being phased out, these agents may still be used for recharging or found in fire extinguishers still in service. The most common halogenated agents are Halon 1301 and Halon 1211. They are often referred to as clean agents because they leave no residue. Because they leave no residue, halon extinguishers are commonly used in areas where sensitive electronic instruments and equipment, including computers, are located. Halon is a very effective fire-extinguishing agent. It is a colorless gas and is very expensive. Though relatively nontoxic if exposure times are limited, it was discovered that halon, which is a fluorocarbon, is thought to cause breakdown of the Earth's protective ozone layer. As a result, halon is being phased out as an extinguishing agent. It can still be found in portable fire extinguishers and fixed extinguishing systems but can no longer be manufactured in the United States. Halon 1301 is generally a nontoxic gaseous fire extinguishing agent and a member of the alkyl halide hydrocarbon derivative family with a chemical formula of CF3Br. The chemical name for

Bromotrifluoromethane– CBrF₃

$$F-\overset{\displaystyle F}{\underset{\displaystyle F}{C}}-Br$$

FIGURE 6.16

Halon 1301 is bromotrifluoromethane and it is also known by the trade names BTM and Freon 1301. Figure 6.16 shows the structural and molecular formula for Halon 1301, which is a member of the hydrocarbon derivative family of chemicals called alkyl halides. They are made up of carbon and one or more of the halogens, fluorine, chlorine, bromine, and iodine.

Figure 6.17 is an illustration of the process used to develop the halon numbering system. Numbers for halon agents are based upon the chemicals present in the formulation, which includes carbon and the first four members of Family VII on the periodic table of the elements. Family VII is composed of fluorine, chlorine, bromine, and iodine. The first digit of a halon number is the number of carbon atoms in the formulation. Second is the number of fluorine atoms, and the third digit is the number of chlorine atoms. The final digit is the number of bromine atoms. If a fifth number is present, it represents the number of iodine atoms. Using that information, Halon 1301 would have one atom of fluorine, three atoms of chlorine, no atoms of bromine, and one atom of iodine.

Halon 1301 is a very effective extinguishing agent for flammable liquid fires. Halon 1301 is designed to be used for fires where there are electrical hazards, telecommunications equipment, flammable and combustible liquids and gases, and other high-value assets such as computers and related equipment. Halon 1301 should not be used on certain chemicals because of potential adverse reactions. These include mixtures of cellulose nitrate and gunpowder; reactive metals such as sodium, potassium, magnesium, titanium, zirconium, uranium, and plutonium; metal hydrides; and chemicals capable of undergoing auto-thermal decomposition (polymerization), such as organic peroxides and hydrazine. In order for halon extinguishing systems to be effective, they must be installed in an enclosed environment capable of containing the gas to maintain the proper percentage of agent in the room for extinguishment

Naming of Halons

1	2	1	1
One Carbon	Two Fluorines	One Chlorine	One Bromine
1	3	0	1
One Carbon	Three Fluorines	No Chlorine	One Bromine

FIGURE 6.17

when discharged. Halon 1301 has an operational temperature range of −70 to 900°F. While generally nontoxic, unnecessary exposure to Halon 1301 should be avoided.

Halon 1211 also known as bromochlorodifluoromethane, BCF (figure 6.18), or Freon 1211. Halon 1211 was introduced in 1973 for protection of highly valuable electronic equipment and other materials in museums, mainframe rooms, and telecommunication switching centers. Halon 1211 is much more toxic than Halon 1301, and firefighters should wear SCBA when entering atmospheres of Halon 1211. It is nonconductive and has lower toxicity than carbon tetrachloride, which was one of the first halons in use for many years. However, carbon tetrachloride was discontinued as a fire extinguishing agent in the 1960s because it was discovered that it formed phosgene in contact with heat. Phosgene is highly toxic and can be harmful to people using the fire extinguishers.

Bromochlorodifluoromethane– CF_2ClBr

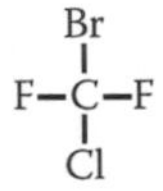

$$F-\overset{\overset{\displaystyle Br}{|}}{\underset{\underset{\displaystyle Cl}{|}}{C}}-F$$

FIGURE 6.18 Structural and Molecular Formula for Halon 1211.

MONTREAL PROTOCOL

The Montreal Protocol on Substances That Deplete the Ozone Layer is an international treaty designed to protect the ozone layer by phasing out the production of a number of substances believed to be responsible for ozone depletion. This treaty was put in force in January 1989. The treaty is centered on several groups of halogenated hydrocarbons that have been shown to play a role in ozone depletion. All of these ozone-depleting substances contain either chlorine or bromine (substances containing only fluorine do not harm the ozone layer). Total phase-out of the suspected substances must be completed by 2030.

Halocarbon (Clean) Agents

Because of the stratospheric ozone-depleting characteristics of halon agents, replacements for halon have been developed. These are sometimes referred to as "clean agents." Clean agents are defined as those that vaporize readily and leave no residue. Because of their environmental impacts, halogenated agents are slowly being replaced with halocarbons, which are thought to be safe for the environment. Clean agents are chemical fire-extinguishing agents developed in response to international restrictions on the production of certain halon fire-extinguishing agents under the Montreal Protocol signed September 16, 1987, as amended. If the fuel is electrical equipment (Class C fuels), such as computer or telecommunications equipment, virtually all of the clean agents are listed and approved for such hazards when electric power to the equipment is shut down upon discharge of the clean agent, which changes the fire to Class A fuels once the electrical equipment is deenergized. There are currently several commonly used clean agent replacements for halons, including Intergen (IG-541), FM200 (HFC-227ea), Agrotec (IG-01), and HFC-23 (FE13). Intergen is a blend of nitrogen, argon, and carbon dioxide and acts by reducing the oxygen level of a space, making it inert to combustion. FM 200, with a chemical name of heptafluoropropane, CF_3CHF_2, acts by inhibiting combustion. Agrotec (IG-01) is the inert gas argon and acts by reducing oxygen. HFC-23 (FE13), trifluoromethane or CHF$_3$, acts by inhibiting combustion.

OBSOLETE FIRE EXTINGUISHERS AND AGENTS

Portable fire extinguishers have been around for over one hundred years. During that time, several extinguishing agents were developed or containers used that ultimately became hazards to the operators and are no longer recognized or listed for use. Carbon tetrachloride (CCl$_4$) was one of the first halon extinguishing agents, halon 1040. Chlorobromomethane (CBM) was also used as a fire extinguishing agent. Carbon tetrachloride was used in fire extinguishers of various sizes and configurations. Glass grenades full of carbon tetrachloride were thrown at the base of fires to extinguish them (figure 6.19). It was also used in glass grenades that were hung from ceilings and had fusible links attached to a metal plunger that would break the glass when released. The agent would then be released to extinguish the fire. Carbon tetrachloride was also used in a variety of hand pump extinguishers that have become popular collector's items (figure 6.20; most were made of brass and are very attractive when polished). It was discovered that carbon tetrachloride in contact with heat forms phosgene gas. Phosgene gas is very toxic, and there is plenty of heat in a fire. When the fire extinguisher is discharged on the fire, the carbon tetrachloride reacts with the heat to release the phosgene. Because of the toxicity of phosgene, carbon tetrachloride is no longer used as a fire-extinguishing agent. OSHA specifically prohibits the use of fire extinguishers containing carbon tetrachloride or chlorobromomethane extinguishing agents in the workplace. Soda acid fire extinguishers are another type portable extinguisher that is no longer used. Extinguishing containers were made of copper, brass, and steel (figure 6.21). A glass or plastic jar of an acid, usually sulfuric, was placed into a holder inside the lid of the metal outer container.

FIGURE 6.19 Carbon tetrachloride was used in various sizes and configurations of glass grenades that were thrown at the base of fires.

FIGURE 6.20 Carbon tetrachloride was also used in a variety of hand pump types of extinguishers that have become popular collector's items.

A stopper was placed into the container of acid (figure 6.22). The remainder of the extinguisher container was filled with soda water. When the extinguisher was inverted (turned upside down), the acid escaped from the glass container and mixed with the soda water, creating a chemical reaction that resulted in pressure, expelling the soda water from the extinguisher to extinguish the fire. The gas created by the chemical reactions between soda water and acid was carbon dioxide. The extinguishers were designed to operate at 100 psi. It was discovered that sometimes the gas production occurred faster than the pressure could be released from the container. Containers, which were primarily liquid or atmospheric containers riveted or soft-soldered, would burst from the excess pressure, causing injury and death. Hoses and elbows on the extinguishers could also become blocked, creating internal pressures in excess of 300 psi. OSHA specifically prohibits the use of all soldered or riveted shell self-generating soda acid or self-generating foam or gas cartridge water type portable fire extinguishers that are operated by

FIGURE 6.21 Soda acid fire extinguishers are another type of portable extinguisher that is no longer used. Extinguishing containers were made of copper, brass, and steel.

FIGURE 6.22 A glass or plastic jar of an acid, usually sulfuric, was placed into a holder inside the lid of the metal outer container. A stopper was placed into the glass container. The remainder of the extinguisher container was filled with soda water.

inverting the extinguisher to rupture the cartridge or to initiate an uncontrollable pressure-generating chemical reaction to expel the agent.

USING FIRE EXTINGUISHERS

Before anyone attempts to use a portable fire extinguisher, they should be trained to do so (figure 6.23). Training is not going over to the wall, picking up the fire extinguisher, and reading the label. Training involves a qualified agency or person providing a structured combination classroom and hands-on training class for all users. When a fire is discovered, it is very important that the fire department be called immediately, *before* attempts are made to use portable fire extinguishers. If the fire is still small (trash can size), a portable fire extinguisher can be used to

FIGURE 6.23 Before anyone attempts to use a portable fire extinguisher, they should be trained to do so.

extinguish the fire. If the fire has spread beyond the point of origin, do not attempt to fight the fire; evacuate the building and let the fire department take care of the fire. The primary reason for calling the fire department first is to avoid delays in response, especially if the fire extinguisher does not work or does not have any affect on the fire. It is also important in case something happens during the use of the extinguisher that injures the user. When using a portable fire extinguisher, make sure to keep the exit from the room, area, or building at your back. You always want to make sure you have a way out. Remove the fire extinguisher from the wall or location where it is kept. Read the label to make sure the extinguisher is rated for the hazard involved in the fire; i.e., ordinary combustibles, flammable liquids, energized electricity, combustible metals, or grease fires. If the extinguisher is not rated for the hazard, do not attempt to use it. Doing so can make the fire worse, spread the fire, or cause injury to the user. Remember to use the "buddy system"; after all, two persons are better than one, especially if something should go wrong. If the extinguisher is the proper type, check the gauge (if one is present) to make sure it is in the green operational range (carbon dioxide fire extinguishers do not have gauges). Briefly discharge the carbon dioxide to make sure it is operational before attacking the fire. Also check the nozzle of any extinguisher to make sure it is not obstructed. Obstructions in the nozzle not only keep the extinguisher from discharging properly, they can result in rupture of the hose and/or container, which can injure the user. Once the extinguisher has been checked, pull the pin, squeeze the trigger, and aim the extinguisher at the base of the fire. Keep low to avoid any smoke that may be present. Stay approximately ten feet away from the base of the fire. Using a sweeping action, move the extinguisher nozzle back and forth over the base of the fire. If any reactions occur or the fire does not seem to be affected by the extinguisher, immediately evacuate the area and wait for the fire department to arrive.

When pressurized foam extinguishers are used, foam should not be discharged directly at the base of the fire as with other portable extinguishers. Foam should be banked off other surfaces down onto the fire. If it is not possible to bank the foam, then the user should stand back far enough to allow the foam to fall lightly on the surface of the burning liquid. Often flammable liquid fires involve tanks, containers, or other metal or nonmetal surfaces that can become hot during a fire. Foam extinguishers function by excluding oxygen from the fire. Care should be taken that flashback does not occur following extinguishment. If the foam blanket breaks down or is disrupted for any reason, oxygen can once again be exposed to the fuel. Hot surfaces can serve as an ignition source, and reignition can occur. If polar solvent types of flammable liquids are encountered, only alcohol or polar solvent types of foam will be effective in extinguishment. Make sure the foam to be used is rated for the type of fire encountered.

TRAINING

NFPA and OSHA, along with local authorities having jurisdiction, have various rules and regulations concerning use of portable fire extinguishers. OSHA requires that where portable fire extinguishers have been provided for employee use in the workplace the employees are trained to familiarize them with the general principles of fire extinguisher use and the hazards involved with incipient-stage firefighting. Training is required to be provided on initial employment and annually thereafter. The bottom line is that with proper training and the proper type of fire extinguisher, and when used properly, small fires can be effectively extinguished. However, if people are not trained to use portable fire extinguishers, they should not use them. Many agencies, including the local fire department, fire extinguisher companies, and state and local training agencies provide portable fire extinguisher training. Some of the training involves classroom sessions covering the chemistry of fire, classes of fire, ratings of fire extinguishers, types of agents, and selection of extinguishers. No fire extinguisher training can be complete with out hands-on training allowing each student to use fire extinguishers on a live fire. Demonstrations should also be presented to show how some agents are ineffective on certain types of fires. Fire props can include pans of varying sizes and depths with flammable liquids on fire, trash can–size fires using ordinary combustibles, and various sized props using propane gas as a fuel. Propane gas props require the least amount of cleanup and are the most environmentally friendly. They are also good for the environment in that they do not release any harmful air pollutants like flammable liquid fires do.

Propane burners can be purchased commercially or can be built by enterprising users. The University of Arkansas Environmental Health & Safety (EHS) department developed a training prop for fire extinguishers that uses propane as a fuel. For further information about the training prop, contact the University of Arkansas EHS department. At the University of Maryland, Baltimore, our plumbing shop constructed the device from plans provided from Arkansas (figure 6.24). It turned out very well and has been used many times successfully for extinguisher training on campus. Components of the training prop are a thirty-gallon barrel-type of container

FIGURE 6.24 Propane burners can be purchased commercially or can be built by enterprising users. The University of Arkansas Environmental Health & Safety (EHS) department has developed a training prop for fire extinguishers that uses propane.

with the top open. A propane burner is constructed in the bottom with valves and connections to a twenty-pound propane tank. When ignited, the burner sends flames several feet above the container, simulating a trash can fire. The only thing we have to sweat is that no one contacts the Baltimore City Fire Department, as they have a ban on open burning in the city and the training prop may violate that ban! I did not ask, because sometimes it's better to ask for forgiveness than permission!

7 Elevators and Controls

INTRODUCTION TO ELEVATORS

Development of the passenger elevator in 1854 by Elisha Graves Otis in Yonkers, New York, made it possible for the creation of high-rise buildings. There are presently an estimated six hundred thousand elevators in the United States and 120 billion rides on elevators each year. A large number of fire and rescue personnel respond to emergencies on these elevators. The subject needs to be addressed throughout the fire service to better prepare those who are called to rescue someone out of an elevator during an emergency or disaster. Today elevators are present in most buildings two stories or higher throughout the United States and the rest of the world. Since firefighters respond to emergencies in all types of buildings, including people "trapped" in elevators, firefighters should have a good working knowledge of elevator operation. According to the American Society of Mechanical Engineers' document ASME A17.1, an elevator is defined as "a hoisting and lowering mechanism, equipped with a car or platform, which moves in guide rails and serves two or more landings." Generally, the purpose of elevators is to carry people and materials vertically from one floor in a building to another. Elevators in most jurisdictions are not part of the means of egress for emergency evacuation even though the codes do allow elevators for emergency use if certain conditions are present (figure 7.1). Firefighters and other emergency personnel may use elevators during emergencies by activating the fire service function of the elevator (figure 7.2). Elevator installation and maintenance is governed primarily by ASME A17.1, *Safety Code for Elevators and Escalators*, with references also in NFPA fire codes and model building codes for certain situations. Specifically, the *International Building Code* addresses elevators in several sections in the index for particular subject matter. NFPA 101, the *Life Safety Code*, under "Means of Egress," "Features of Fire Protection," and each of the occupancy chapters, contains elevator information. NFPA 13, the *Sprinkler Code*, covers sprinkler requirements; section 8, article 620 of NFPA 70, the *National Electrical Code*, lists electrical requirements for elevator installations; and section 3 of NFPA 72 cites fire alarm issues that apply to elevators.

During fires, fire inspectors and firefighters should be careful when entering elevator machine rooms. There are moving parts such as cables and drive sheaves that can injure or kill you. Elevators operate under high voltage. If elevator machine rooms are sprinkled, there may be water coming from sprinkler heads or water standing on the floor. You do not want to be standing in water when you grab the main disconnect switch with your unprotected hand. Access to the rooms should be kept clear for easy access under emergency conditions. Nothing should be stored in the elevator machine room unless it is directly connected with the elevator. Keys for elevator operation should be made available to responding firefighters to make use of

FIGURE 7.1 Elevators in most jurisdictions are not part of the means of egress for emergency evacuation even though the codes do allow elevators for emergency use if certain conditions are present.

Phase I and Phase II elevator fire service operation. Keys should also be available for the elevator machine room in the event that there is a fire in that location. Before any rescue attempts are made involving an elevator, firefighters should go to the elevator machine room and shut off the elevator main disconnect switch. Always follow the information provided in ASME A17.4 when performing elevator rescues or operating elevators during fires.

FIGURE 7.2 Firefighters and other emergency personnel may use elevators during emergencies by activating the fire service function of the elevator.

ELEVATOR NOMENCLATURE

- Elevator Hoistway: an opening through a building or structure for the travel of elevators extending from the pit floor to the roof or floor above.
- Elevator Pit: the portion of the hoistway extending from the sill level of the lowest landing to the floor at the bottom of the hoistway.
- Shunt-Trip: shutdown of electric power to the elevator resulting from heat detector activation in the hoistway or elevator machine room prior to sprinkler operation.
- Firefighter Service: function that allows firefighters to take control of elevators through key access to use them as needed for firefighting and rescue operations.
- Recall: return of the elevators to predesignated floors when smoke detectors located in elevator lobbies are activated.
- Machine Room: location of elevator equipment and controls. Usually located on floors above the top of the elevator shaft.

TYPES OF ELEVATORS

HYDRAULIC ELEVATORS

Hydraulic elevators are powered by energy applied by means of a liquid under pressure in a hydraulic jack (figure 7.3). Hydraulic elevators are also operated by hydraulic driving machines or roped hydraulic driving machines. Liquids used in hydraulic elevators are generally hydraulic fluid specially made for use in hydraulic elevator systems. Hydraulic fluids may either be flammable or nonflammable. Use of nonflammable hydraulic fluids results in fewer code requirements in terms of sprinklers

FIGURE 7.3 Hydraulic elevators are powered by energy applied by means of a liquid under pressure in a hydraulic jack.

FIGURE 7.4 A tank (the fluid reservoir) for a hydraulic elevator.

and detection devices in elevator pits. Hydraulic elevator systems lift a car using a hydraulic ram (a fluid-driven piston mounted inside a cylinder). The cylinder is connected to a fluid-pumping system. The hydraulic system has three parts:

- A tank (the fluid reservoir; figure 7.4)
- A pump, powered by an electric motor (figure 7.5)
- A valve between the cylinder and the reservoir (figure 7.6)

The pump forces fluid from the tank into a pipe leading to the cylinder. When the valve is opened, the pressurized fluid will take the path of least resistance and

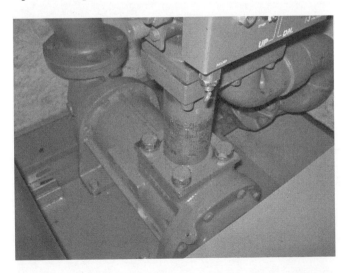

FIGURE 7.5 A pump, powered by an electric motor, for a hydraulic elevator.

FIGURE 7.6 A valve that controls the flow of hydraulic fluid is located between the cylinder and the reservoir for a hydraulic elevator.

return to the fluid reservoir. But when the valve is closed, the pressurized fluid has nowhere to go except into the cylinder. As the fluid collects in the cylinder, it pushes the piston up, lifting the elevator car. When the car approaches the correct floor, the control system sends a signal to the electric motor to gradually shut off the pump. With the pump off, there is no more fluid flowing into the cylinder, but the fluid that is already in the cylinder cannot escape (it cannot flow backward through the pump, and the valve is still closed). The piston rests on the fluid and the car stays where it is. To lower the car, the elevator control system sends a signal to the valve. The valve is operated electrically by a basic solenoid switch. When the solenoid opens the valve, the fluid that has collected in the cylinder can flow out into the fluid reservoir. The weight of the car and the cargo pushes down on the piston, which drives the fluid into the reservoir. The car gradually descends. To stop the car at a lower floor, the control system closes the valve again. Installation of hydraulic elevators is limited to buildings that are five stories or less in height.

TRACTION ELEVATORS

Traction elevators are driven by an electric motor located in the buildings elevator machine room(s). The motor has a pulley or drive sheave attached and there are usually three or more cables on it that are fastened to the elevator. In some older elevators there may be only one cable. Traction elevators have safeties or emergency brakes on them that are supposed to bring the elevator to a stop in case of a free fall. There are two basic types of traction elevators, geared and gearless. Geared elevators are operated by geared traction machines driven by electric motors. Worm gears mechanically control car movement by coiling a steel hoist rope around a grooved wheel that is attached to a gear box operated by a high-speed motor (figure 7.7).

FIGURE 7.7 Worm gears mechanically control car movement by coiling a steel hoist rope around a grooved wheel, which is attached to a gear box operated by a high-speed motor.

Geared elevators are generally used for elevators with speeds of up to 1,000 ft/min (5 m/s). Gearless traction machines are high-speed and low-torque electric motors (figure 7.8). Grooved wheels or pulleys are attached directly to the motor. When the motor turns one way, the elevator car goes up. When it turns the other way, the elevator car goes down. A brake is mounted between the motor and the wheel to hold the car at a particular floor. Brakes are usually an external drum type actuated by spring

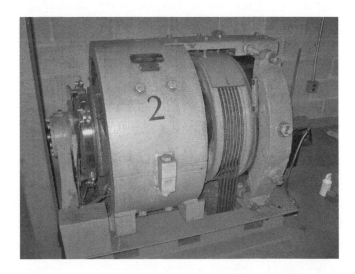

FIGURE 7.8 Geared elevators are generally used for elevators with speeds of up to 1,000 ft/min (5 m/s). Gearless traction machines are high-speed and low-torque electric motors.

FIGURE 7.9 In buildings more than four stories in height, at least one elevator car must be of an appropriate size to accommodate ambulance stretchers.

force and held open electrically. A power failure will cause the brake to engage and prevent the elevator from falling. Regardless of type of traction elevator, cables are attached to the elevator car or may be looped around the car and attached to a counterweight. Grooves in the wheel prevent the ropes from slipping. Traction is provided to the ropes by the groove in the wheel; hence the name "traction elevator."

In buildings more than four stories in height, at least one elevator car must be of an appropriate size to accommodate ambulance stretchers (figure 7.9) and provide for fire department emergency access to all floors (elevator fire service).

ELEVATOR CONSTRUCTION

According to the *International Building Code* (IBC); NFPA 101, the *Life Safety Code*; and ASME 17.1, *Safety Code for Elevators and Escalators*, elevator hoistways, or shafts as they are sometimes called, are required to be constructed of materials that provide a two-hour fire resistance rating (figure 7.10). No more than four elevators can be included in any one fire-rated hoistway enclosure. Elevators and fire-rated stairwells are not allowed to be within the same shaft enclosure. No plumbing or mechanical systems are allowed to be in the elevator hoistway. The elevator shaft cannot be used as a mechanical shaft for other building components. Within buildings in which corridor walls carry a fire resistance rating, elevator lobbies must be provided with an enclosure separating it from the corridor with an appropriate fire protection rating. There are some exceptions to the lobby enclosure requirement, such as a complete automatic sprinkler system installed throughout the building. Sprinkler systems may also allow for other code trade-offs when they are installed. Consult the building code for the specifics of the exceptions and any other requirements.

FIGURE 7.10 Elevator hoistways, or "shafts" as they are sometimes called, are required to be constructed of materials that provide a two-hour fire resistance rating.

Hoistways in buildings constructed with three stories or higher are required to have smoke venting of the hoistway for the removal of smoke and hot gases during a fire condition. With the exception of building occupancy Groups R-1 (residential, boarding houses, hotels, and motels transient), R-2 (residential, apartment houses, boarding houses not transient, convents, dormitories, fraternities and sororities, monasteries, vacation timeshare properties, and motels and hotels nontransient), I-1 (institutional residential board and care facilities, assisted-living facilities, halfway houses, group homes, congregate care facilities, social rehabilitation facilities, alcohol and drug centers, and convalescent facilities), and I-2 (institutional, hospitals, nursing homes, mental hospitals, and detoxification facilities), the smoke venting may not be required in a fully sprinkled building. NFPA 13 exempts elevator hoistways from sprinkler coverage. This is allowed because of the two-hour separation, noncombustible shaft construction, and noncombustible elevator cars. This allows the building to be considered fully sprinkled even though the hoistway is not. In addition to elevator hoistways, the IBC exempts certain locations within a building from sprinkler coverage while still considering the building fully sprinkled. They include rooms or areas that are of noncombustible construction with wholly noncombustible contents. Therefore, if the elevator shaft is of noncombustible construction and the elevator cars are noncombustible, sprinklers are not required in the shaft, and smoke venting is not required in a fully sprinkled building. Now having said that, be prepared to sell your case to the elevator inspector. There are often conflicts between the elevator and fire codes. Elevator inspectors may interpret the elevator code differently than the fire inspector and they may also have a different interpretation of the fire codes, although you would think they would leave fire code interpretation to the fire inspector.

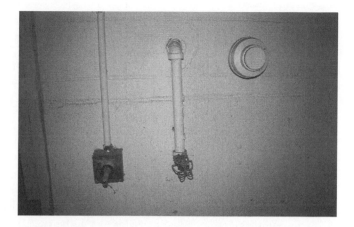

FIGURE 7.11 When sprinklers are installed in hoistways, heat detectors have to be installed to activate prior to the sprinkler heads and activate the shunt-trip function of the elevator controls.

The advantage of not sprinkling hoistways is that if there are no sprinklers, alarm detection is not required at the top of the hoistway, which removes the requirement for shunt-tripping from the hoistway as well. Shunt-tripping will be discussed in detail later in this chapter. When sprinklers are installed in hoistways, heat detectors have to be installed to activate prior to the sprinkler heads and activate the shunt-trip function of the elevator controls (figure 7.11). Shunt-tripping shuts off all power to the elevator prior to sprinkler activation. Having smoke detectors in elevator hoistways or elevator pits places them in situations where they will be subject to false alarm (figure 7.12). They are also difficult to test and maintain. NFPA 72 prohibits smoke detectors from being placed at the top of elevator hoistways, which is

FIGURE 7.12 Having smoke detectors in elevator hoistways or elevator pits places them in situations where they will be subject to false alarm.

in direct conflict with the elevator code. Sprinkler coverage is required in elevator pits for hydraulic elevators that use flammable hydraulic fluid. If the elevator is not hydraulic or nonflammable fluid is used, then no sprinkler heads and no detection devices are required in the pit. Smoke detectors in elevator hoistways are subject to false alarm because of the potentially dusty atmosphere. Constant movement of the elevator car(s) stirs up any dust that may be in the shaft.

ELEVATOR MACHINE ROOM

Elevator machine rooms may be found at the top of the elevator hoistway, in a basement mechanical room, or on one of the building's occupied floors (figure 7.13). Only equipment and machinery associated with the operation of the elevators is allowed in the elevator machine room. Only such electrical wiring, raceways, and cables used directly in connection with the elevator are allowed to be in the hoistway and machine rooms. These include wiring for signals, communication with the car, lighting, heating, air-conditioning, and pit sump pumps. Machine rooms must be constructed to the same or greater rating as the elevator hoistway. Plumbing systems other than sprinkler piping cannot be located in elevator machine rooms because of all of the electrical equipment located there. Elevator machine rooms should be restricted areas, and only elevator maintenance personnel or elevator companies should be allowed access. Machine rooms should not be used for storage of any kind. ASME A17.1 requires that an ABC-rated portable fire extinguisher be located in all elevator machine rooms. You may still find CO_2 fire extinguishers in some machine rooms because the requirement for ABC is still relatively new. NFPA 101, the *Life Safety Code*, requires elevator machine rooms with a travel distance exceeding fifty feet (15 m) above the level of exit discharge or thirty feet (60 m) below the level of exit discharge be provided with independent ventilation or air-conditioning systems to

FIGURE 7.13 Elevator machine rooms may be found at the top of the elevator hoistway, in a basement mechanical room area, or on one of the building's occupied floors.

maintain temperatures during firefighter emergency operations. If emergency power is connected to elevators, then the ventilation or air-conditioning systems also need to be on emergency power. Elevator machine rooms are usually sprinkled. Because of the electrical equipment in the room, having water discharge from the sprinkler system can be damaging to the equipment. When installing sprinklers in the machine room, it is a good idea to have a shutoff valve in the sprinkler piping just outside the machine room so that the water can be shutoff as soon as the fire is extinguished. This action will reduce water damage to electrical equipment and flooding in other parts of the building. There will also need to be a tamper switch installed in association with the valve and hooked into the building fire alarm system.

ELEVATOR RECALL

Fire alarm initiating devices (smoke or heat detectors) are installed at each elevator floor (figure 7.19) in the elevator lobby. Heat detectors are used at locations where smoke detectors would be subject to false alarms (mechanical rooms, parking

FIGURE 7.14 Elevators are provided with fire service control features to assist firefighters in taking control of and using the elevator during firefighting operations or rescue of handicapped occupants.

FIGURE 7.15 Elevator phase status should be marked at some location on the elevator door frame. If there is no fire service, then the sign should state NO FIRE SERVICE.

garages, and areas where steam is present, to mention a few; figure 7.20). When steam is present that may affect heat detector operation, activation temperatures should be set above 212°F. Activation of any of the elevator lobby smoke or heat detectors will cause the elevators in that bank to recall automatically non-stop to the predesignated floor (figure 7.21). If the fire alarm device that activates is located on the predesignated floor, then the elevators in that bank are recalled to a predetermined

FIGURE 7.16 If the Elevators have Phase I or Phase II then the sign should state which Phase the elevator is equipped with.

alternate floor. That alternate floor should not be a floor below the level of exit discharge from the building if at all possible. Once recalled to the predesignated floor, elevators will be locked out of service until placed in fire service by responding firefighters or the elevators are reset. Initiating devices such as smoke detectors, heat detectors, and sprinkler flow switches in or associated with elevator hoistways and elevator machine rooms will also activate elevator recall. NFPA 72, the *National Fire Alarm Code*, allows for smoke and heat detectors in elevator hoistways and machine rooms to initiate a supervisory signal rather than an alarm signal. Because the devices are actually installed because of the elevators and not required by the fire alarm code, this does not create any life safety concerns for occupants. Placing smoke and heat detectors in hoistways and machine rooms on supervisory signal greatly reduces the chances for false alarms from these devices. Supervisory signals from a fire alarm system are required by the code to be investigated within one hour of activation.

ELEVATOR FIRE SERVICE FUNCTION

First of all, building occupants should not use elevators during a fire or when the building fire alarm is sounding. Elevators can be programmed to be recalled automatically when the fire alarm system in the building is activated, which will keep building occupants from using elevators during a fire. Elevators are generally not safe for civilians to use under fire conditions. Firefighters, however, may be able to use elevators to transport personnel and equipment if it is determined that it can be done safely. Elevators are provided with fire service control features to assist firefighters in taking control of and using the elevator during firefighting operations or rescue of handicapped occupants (figure 7.14). ASME A17.1, *Safety Code for Elevators and Escalators*, details the two phases of emergency elevator operation. Elevators can actually be found with three types of firefighter service situations on new and existing installations (the third type not mentioned in ASME is elevators without any fire service available at all). It should be noted that not all elevators have fire service controls (buildings four stories or less do not require them). Older buildings with older elevators may not have a fire service function. When elevators do not have fire service controls, firefighters will have no way to recall elevators or take them out of automatic operation. They will not be able to operate the elevators safely under fire conditions. It is not recommended that firefighters use elevators without firefighter service controls. Elevators may go to the fire floor and open the doors. If smoke and fire are present, firefighters may become trapped because they will be unable to control the elevator. All new elevator installations are required to be installed with both Phase I and Phase II firefighter service. Elevator phase status should be marked at some location on the elevator doorframe. If there is no fire service, then the sign should state "No Fire Service" (figure 7.15). If the elevators have Phase I or Phase II service, then the sign should state which phase the elevator is equipped with (figure 7.16). Marking elevators in this manner allows firefighters and other emergency responders to know what type of firefighter service, if any, is associated with a particular elevator.

Phase I firefighter service consists of a recall system that automatically sends elevators to a designated primary level. This phase allows the firefighters to use an elevator key to recall the elevator(s) to the designated floor (usually the first or

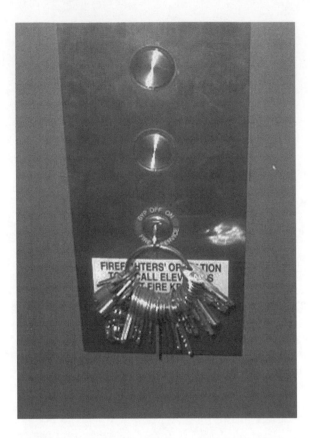

FIGURE 7.17 Each Phase I elevator bank will have a three-position key slot located in the elevator lobby within sight of the elevator(s) at the main floor (usually ground or first floor).

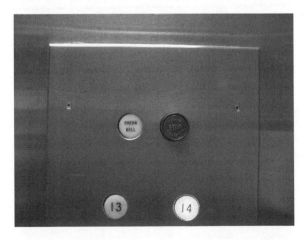

FIGURE 7.18 If an elevator has been held at a location with the emergency stop switch located in the car, it will remain at that location until the switch is released.

FIGURE 7.19 Fire alarm initiating devices (smoke or heat detectors) are installed at each elevator floor.

ground floor). Each Phase I elevator bank will have a three-position key slot located in the elevator lobby within sight of the elevator(s) at the main floor (usually ground or first floor; figure 7.20). The key slot switch cannot be located behind a locked door or cover. The three-position switch should be marked with "Bypass," "Off," and "On" (in that order). Only Phase I switches or fire alarm initiating devices at elevator floors, machine rooms, or hoistways are allowed to initiate Phase I operation. When the Phase I switch is in the "Off" position, elevators will operate normally and be controlled by the smoke or heat detectors in the lobbies. When the Phase I switch is in the "Bypass" position, normal elevator operations are restored regardless of the status of elevator lobby detectors. When the Phase I switch is in the "On" position, all elevator cars controlled by the switch (usually only those in the elevator bank) will return to the designated level non-stop and doors will open and remain open. Any car traveling in the opposite direction at the time of the Phase I recall will reverse directions without going to the next level and without opening the doors and return to the designated floor and open the doors. If an elevator has been held at a location with the emergency stop switch located in the car, it will remain at that location until the switch is released (figure 7.21). When released, the emergency stop will be

FIGURE 7.20 Heat detectors are used at locations where smoke detectors would be subject to false alarms (mechanical rooms, parking garages and areas where steam is present to mention a few).

inoperative and the car will return non-stop to the predesignated floor. When Phase I fire service has been activated, a fire hat (helmet) symbol in the elevator car will light up. This is an indication that the elevator is in fire service Phase I. The fire hat will remain illuminated until the elevator is placed back into automatic operation. When in "Bypass" mode, all elevator call buttons are required to be inoperative. All call lights' directional lanterns will be extinguished and remain inoperative. If used, car position indicators will also be extinguished and remain inoperative while in

FIGURE 7.21 Activation of any of the elevator lobby smoke or heat detectors will cause the elevators in that bank to return automatically non-stop to the pre-designated floor.

FIGURE 7.22 Phase II firefighter service permits firefighters to manually take control of the elevators with an override key.

"Bypass." When an initiating device is activated in the hoistway or elevator machine room, the fire hat symbol in the elevator car that was illuminated by activation of Phase I fire service will flash intermittently in elevators with equipment in that hoistway or machine room. The intermittent illumination of this fire hat symbol alerts firefighters that there is a problem in the shaft or machine room and they should exit the elevator as soon as possible.

Phase II firefighter service permits firefighters to manually take control of the elevators with an override key (figure 7.22). Elevators with Phase II operation also have Phase I elevator recall as well. Phase II overrides all automatic controls, including the Phase I recall. The elevator key places the elevator in firefighter service and the firefighters will then take the key into the elevator car to operate the controls within the car. One very important feature of a Phase II elevator operation is that it allows firefighters to override the closing controls on an elevator door and close the door manually in an emergency. When Phase II control is provided on an elevator, a three-position key-operated switch is installed in the elevator car. This switch is only operational when the Phase I switch outside the elevator car in the lobby has been turned on or the elevator has been recalled by a lobby detection device. Phase II key switches are required to rotate clockwise from "Off" to "Hold" to "On" (figure 7.23). The key can be removed from the switch at any of the three settings. It is recommended that firefighters not leave keys in the elevator car while the car is in fire service to insure that no one else will try to use the car. Phase II operation in the elevator is only activated when the three-position switch is in the "On" position. Elevators should only be used in Phase II operation by trained emergency service personnel. Under Phase II operation, the elevator can only be operated by a person in the car. Door open and close buttons need to be operated manually and floor call buttons

FIGURE 7.23 Phase II key switches are required to rotate clockwise from "Off" to "Hold" to "On."

in the car need to be activated or the car will not move. All lobby call buttons and directional lanterns remain inoperable. Floor indicators should remain operational. Opening and closing of car doors is accomplished by continuous pressure on the open or close button. Elevator cars are required to have buttons installed and marked "Call Cancel" (figure 7.24). These buttons are only operational under Phase II operation.

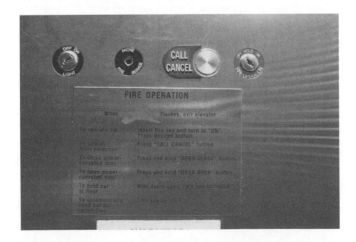

FIGURE 7.24 Elevator cars are required to have buttons installed and marked "Call Cancel."

FIGURE 7.25 Elevators may also be equipped with operating devices on top of the car for maintenance and service known as "inspection operation."

When activated, all floor selections are canceled and the traveling car will stop at the next available landing. When the "Hold" position is activated, the elevator is in Phase II operation. The car will remain at the landing with the doors open. The door close button will be inoperative. When the key position is turned to "Off," the elevator will revert to Phase I operation and be recalled to the designated floor.

Elevators may also be equipped with operating devices on top of the car for maintenance and service known as "inspection operation" (figure 7.25). When an elevator equipped with firefighter service is on inspection operation, a continuous signal that is audible on top of the car activates when the Phase I switch is in the "On" position or when the detection device is activated to alert the operator of an emergency. The elevator car remains under the control of the operator until removed from inspection operation. Inspection operation controls override the Phase I and Phase II firefighters service on the elevator car.

Phase I elevator operating instructions are required to be posted next to the Phase I key switch and Phase II instructions are in the elevator car near the Phase II key switch. Phase I instructions are fairly simple. They need to have a sign with "Firefighters' Operation" as a heading and the instructions "To recall elevators insert fire key and turn to 'On'" (figure 7.26). Phase II operational instructions are more extensive (figure 7.27). All firefighters should be thoroughly familiar with the Phase I and Phase II operations of elevators. It would be helpful if this training was made available during firefighter basic training classes. Departments should develop standard

FIGURE 7.26 Phase I instructions are fairly simple. Elevators should have a sign with "Firefighters Operation" as a heading and the instructions "To recall elevators insert fire key and turn to 'On.'"

operating procedures (SOPs) for use of elevators during fires. As an example of the types of SOPs that should be developed, the Houston, Texas, Fire Department elevator response SOPs are shown below.

- Unless information from dispatch indicates that an elevator occupant is suffering from a medical emergency, calls for stuck elevators shall be handled non-emergency.
- The first arriving unit shall establish scene control and coordinate the rescue or removal effort.
- Elevator rescue or removal operations shall be conducted in such a way as to ensure that no additional risk is placed on occupants or firefighters.
- If unable to perform elevator evacuation without undue risk, wait for the elevator maintenance company to arrive while maintaining constant communications with the occupant(s).

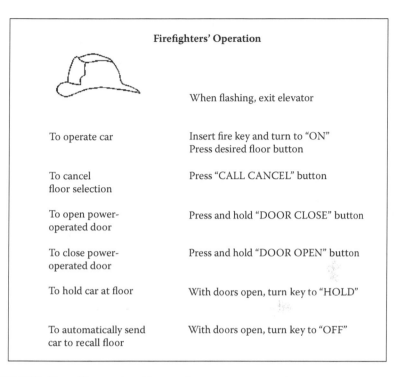

Firefighters' Operation

	When flashing, exit elevator
To operate car	Insert fire key and turn to "ON" Press desired floor button
To cancel floor selection	Press "CALL CANCEL" button
To open power- operated door	Press and hold "DOOR CLOSE" button
To close power- operated door	Press and hold "DOOR OPEN" button
To hold car at floor	With doors open, turn key to "HOLD"
To automatically send car to recall floor	With doors open, turn key to "OFF"

FIGURE 7.27 Phase II operational instructions are more extensive.

- If mechanical entrapment is found, request additional resources.
- The first arriving unit shall confirm that an elevator is stuck, the floor it is located on, the number of occupants, and whether there are medical emergencies involved. If the location of the stuck elevator is not known, it may be necessary to open the elevator shaft door and make a visual inspection.
- If the building is equipped with firefighters service, attempt to recall the elevators(s). In addition, attempt to obtain keys for the elevator mechanical room.
- Assign personnel to identify floors where elevator doors have keyholes.
- Once the elevator is reached, establish contact with the occupants and advise them of the situation. The need for additional resources should be reassessed at this time. Consider contacting the elevator maintenance company and get an estimated time of arrival.
- Assign personnel to the elevator control room to shut down power. Completion of this task shall be reported by radio.
- After confirming that the elevator power is shut down, access the elevator car by keyway. Personnel operating in the vicinity of an open hoistway shall be secured by a tag line.

- Use a ladder to assist an occupant's exit from the elevator whenever the car floor is not even with the exit platform.
- Notify a building representative to have the elevator serviced.

ELEVATOR EMERGENCY POWER

Emergency power, where required or installed, must be manually transferable to all elevators in each bank. If there is only one elevator, the transfer of power shall be automatic within sixty seconds of the main power outage. Where there is more than one elevator or elevator bank on emergency power, the transfer of power shall occur automatically within sixty seconds where the standby power source is of sufficient capacity to operate all elevators at the same time. Where the standby power source is not of sufficient capacity to operate all elevators at the same time, all elevators shall transfer to standby power in sequence, return to the designated landing, and disconnect from the standby power source.

After all elevators have been returned to the designated level, at least one elevator shall remain operable from the standby power source. If elevators are part of an accessible means of egress, they must be provided with standby power at all times. At least one elevator in a building should also be provided with emergency power so the elevator may be operated during power outages to a building. When there is more than one elevator in a building and only one at a time can be on emergency power, a manual switch must be provided to allow firefighters to designate which elevator will operate on emergency power (figure 7.28). This switch can only remove power from an elevator that has stopped. Ideally, when a building has an emergency

FIGURE 7.28 When there is more than one elevator in a building and only one at a time can be on emergency power, a manual switch must be provided to allow firefighters to designate which elevator will operate on emergency power.

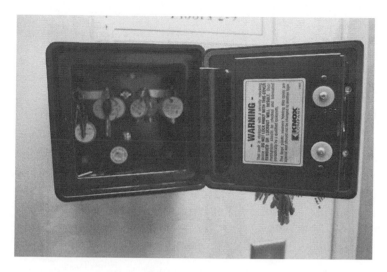

FIGURE 7.29 Rapid Entry Systems can be used to store elevator keys; firefighters generally have keys for the Knox® Boxes in their community.

generator, it should be sized so that all elevators can be operated simultaneously on emergency power. This allows firefighters and emergency responders to operate all elevators if needed during an emergency when the main building power has failed or been shutoff for tactical purposes.

When normal power is restored following a power interruption, the elevator should not be removed from Phase I or Phase II operation. Individual keys should be provided to firefighters for each elevator in a building at a mutually acceptable location. Knox® Box Rapid Entry Systems can be used to store elevator keys as firefighters generally have keys for the Knox® Boxes in their community (figure 7.29). However, other means of access to keys can be utilized as well. Recall of the elevators in a building may also be tied into the operation of the fire alarm system and is highly recommended. When the fire alarm is activated, all building elevators will return to the ground floor and open. This step has the advantages of keeping building occupants from using elevators during fire emergencies or fire alarm activations and the elevators are immediately available to the fire department when they arrive on scene. Elevators with no fire service or Phase I elevators will not allow firefighters to control the opening and closing of elevator doors. Generally, if firefighters choose to use a Phase I elevator or a non–fire service elevator, they will go to the floor below the fire and then use the stairs to access the fire floor.

Solid-state elevator control equipment operates correctly only if maintained within a certain temperature range. NFPA 101, NFPA 5000, and the IBC require independent ventilation in machine rooms containing solid-state equipment that controls elevators traveling over certain distances. Whenever such elevators receive emergency power, their corresponding machine room ventilation would also receive emergency power. These features help maintain at least one elevator operational throughout fire suppression operations.

BATTERY LOWERING UNITS

There are other options available to provide emergency power to get occupants off of elevators in the event of a power failure. A battery lowering unit provides automatic backup power for a hydraulic elevator when the building loses power. The unit returns the elevator to the designated landing and opens the doors, allowing passengers to exit safely. An emergency power lowering system (EPLS) provides automatic or manual return of a traction elevator to a predetermined landing during a blackout or single-phase electrical condition. When the elevator reaches the designated landing, passengers can open the car doors by pressing the "Door Open" button.

FIRE ALARM INTERFACE

Even though fire alarm devices installed in association with elevators are not required by the fire codes, the devices are allowed to be hooked into the building fire alarm system. They may also be hooked into their own separate system. Smoke detectors are required to be located in the elevator machine room for operation of the firefighter hat function of elevator fire service (figure 7.30). When sprinkler heads are in the machine room, heat detectors are required to be placed next to the sprinkler heads with a temperature setting that is lower than the temperature needed to activate the sprinkler heads (figure 7.31). Smoke and heat detectors in

FIGURE 7.30 Smoke detectors are required to be located in the elevator machine room for operation of the firefighter hat function of elevator fire service.

FIGURE 7.31 When sprinkler heads are in the machine room, heat detectors are required to be placed next to the sprinkler heads with a temperature setting that is lower than the temperature needed to activate the sprinkler heads.

elevator machine rooms may be placed on supervisory rather than alarm signals because their primary use is for elevator operation and not life safety in the building. Any detection devices installed in the hoistway may also be placed on supervisory signal. Doing this does not reduce the life safety of building occupants and does reduce the chance of false alarms.

SAFETY DEVICES

Emergency doors for rescue are required to be provided where an elevator is installed in a single blind hoistway or on the outside of a building. In the blind portion of the hoistway or blank face of the building, an emergency door in accordance with ASME A17.1 must be installed.

SHUNT-TRIP

Currently, ASME A17.1 requires an automatic power shutdown feature for elevators that have fire sprinklers located in their machine rooms and, under certain conditions, in hoistways. Shutdown occurs prior to the discharge of water, usually when heat detectors mounted next to each sprinkler head are activated. These heat detectors have both a lower temperature rating and a higher sensitivity (a lower response time index) than the sprinkler. What happens when the heat detector activates is the immediate power shutdown to the elevator. It does not move to another floor, so anyone in the elevator at the time becomes trapped in the elevator wherever it stops in the shaft. However, to minimize the chance that firefighters will be trapped by a power shutdown, the temperature rating of the heat detector should be as high as feasible. The 2006 edition of ASME A17.1 has addressed the issue of potential entrapment when the shunt-trip activates. New requirements call for the heat detectors used for shunt-trip to initiate Phase I emergency recall operation and delay the removal of power and the release of water to allow the completion of recall. If the elevators are already operating on Phase II emergency in-car operation, the recall will not occur, but shutdown and water release will still be delayed. If the elevator is on Phase II operation, the delay will allow the car to go to the next selected floor. Once the car has stopped at the floor, all calls are canceled and the shunt-trip will activate. As a warning to firefighters, when the heat detector is activated for shunt-trip, the fire hat in the elevator car will flash intermittently. Keep in mind that the new shunt-trip regulations will go a long way toward reducing the chances of entrapment in elevator cars as a result of activation of a shunt-trip but it will not totally eliminate the possibility. Caution should be exercised when using elevators under fire conditions. These new regulations will apply to new elevator installations or upgrades of existing elevators. Remember that existing elevators will not likely have this new feature. Local fire prevention personnel should make an effort to get building owners to upgrade elevators to include the new shunt-trip requirements.

An alternate shunt-trip shutdown method involves water flow detectors; however, these detectors cannot employ a time delay, so designers seldom choose this method. Note that, in many cases, NFPA 13, *Standard for the Installation of Sprinkler Systems*, permits sprinklers to be omitted from these areas. Whatever means is used to disconnect the power supply cannot be automatically reset; it must be done manually. This function is known as a "shunt-trip." This must be accomplished in accordance with NFPA 72, section 3-8.15.

The *National Fire Alarm Code*, NFPA 72, requires that smoke detectors in either the elevator hoistway or the elevator machine rooms trigger separate and distinct visible annunciation at both the fire alarm control unit and the fire alarm annunciator. This alarm notifies firefighters that the elevators are no longer safe to use, and it also provides some warning time prior to the shutdown feature that is required with sprinkler protection. In addition, ASME A17.1 requires a warning light in elevator cabs to flash when an elevator problem is imminent. This warning light is a white button with a red firefighter hat superimposed on the button (figure 7.33).

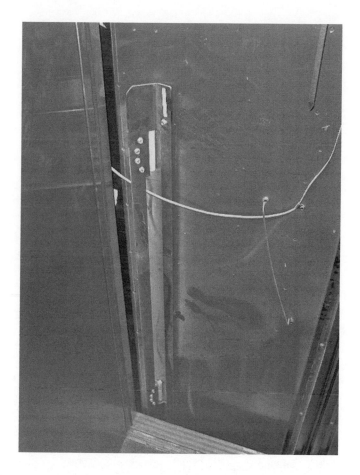

FIGURE 7.32 Door Restrictors prevent the elevator car doors from opening more than four inches when the elevator is not within its "landing zone" (usually 18" above or below the floor landing).

RESCUE OF "TRAPPED" OCCUPANTS

DISCONNECT POWER

Before working on elevators, firefighters should first turn off the power to the elevator at the main switches in the elevator machine room and use ladders or other barriers to block the shaftway opening between the elevator and floor landing. A thorough risk–benefit analysis should be performed before deciding that an elevator rescue operation should be conducted.

ASME A17.4, *Guide for Emergency Personnel*, recommends that anytime an emergency evacuation must take place from an elevator it should be done under the direct supervision of elevator personnel. After all, they are trained and know elevators inside and out and may be able to get access much quicker than emergency response personnel, utilizing elevator controls without damaging equipment. However, elevator personnel

FIGURE 7.33 In addition, ASME A17.1 requires a warning light in elevator cabs to flash when an elevator problem is imminent. This warning light is a white button with a red firefighter hat superimposed on the button.

may not always be readily available. In that case, personnel conducting an emergency rescue should meet the training requirements outlined in ASME A17.4. Most trapped elevator occupants in stalled elevators are safer inside the elevator car awaiting the arrival of an elevator mechanic than being rescued by undertrained firefighters. As long as no one is injured, no one is panicking, occupants have plenty of air, and they are not in danger of fire or hazardous materials, there is no reason or hurry to remove them. Elevator cars generally have telephones or two-way communications systems (figure 7.34) and rescuers can contact car occupants to update them on the progress of the rescue operation. Some facilities have on-site elevator personnel and waiting time for expert assistance is very short. At the University of Maryland, Baltimore, we have our own elevator shop and mechanics respond to all "trapped" persons calls that are reported to the university police or work control. We do not call the fire department unless someone is injured because elevator mechanics on campus can quickly remove persons often before the fire department could have time to respond. Situations where it may take longer for elevator mechanics to arrive will require that decisions be made concerning the removal of occupants by firefighters. If firefighters are not thoroughly familiar with elevator operations, they should wait for the mechanic as long as the occupants are safe inside. Tops of elevator cars are a very dangerous place to be and only trained maintenance personnel should go there. Forcible entry tactics should only be employed on elevator cars during rescues as a last resort because expensive damage can be caused to working parts of elevators. If it's not a life or death situation, wait for the elevator mechanic.

ELEVATOR DOORS

Elevators operate with two sets of doors, one on the car and the other on the opening to the shaft (figure 7.35). However, when the car reaches a given floor, they open

FIGURE 7.34 Elevator cars generally have telephones for emergency communication for problems in the elevator car.

together. Outer elevator doors have a "hole" located near the top of one of the doors. A special key, called an "interlock key," is available that allows the doors to be manually opened to assist in rescue or analyzing the situation to determine tactical rescue options. Car doors are operated by an electric motor controlled by the elevator's computer. As the car doors open, the mechanics of the system also opens the shaft door. Most newer elevators have motion sensors that keep the doors from closing if somebody is between them. Older elevator doors are opened and closed using an electric eye. As long as the beam of the electric eye is broken by someone passing through the door opening, the door will stay open. When the beam is unbroken, the door will close so that the elevator car can move to another floor. Newer elevators utilize a form of heat sensor to keep doors open while passengers are entering or exiting the elevator car.

Door Restrictors

Door restrictors prevent the elevator car doors from opening more than four inches when the elevator is not within its "landing zone" (usually eighteen inches above or below the floor landing; figure 7.32). Elevators installed within the past twenty-five years or so may be equipped with door restrictors. Door restrictors help keep passengers safe by locking the doors when the elevator car is outside the unlocking zone (eighteen inches above and below the landing). Door restrictors have generally been installed on elevators built since 1980 and have been retrofitted on many installed before that time. They were installed in light of accidents in the 1970s and 1980s where people pried open car doors and attempted to exit elevators stopped between floors. Children living in housing projects and students in college dormitories have been known to pry open doors of a moving elevator car, which immediately stops the hoisting machine and sets the brake. This situation then allows occupants to get

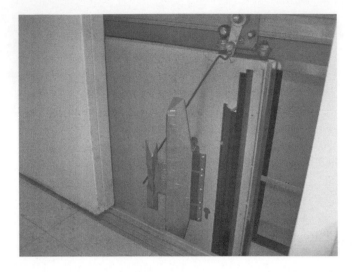

FIGURE 7.35 Elevator doors operate with two sets of doors, one on the car and the other on the opening to the shaft.

on top of an elevator car, sometimes referred to as "elevator surfing." Some of these people were decapitated, lost limbs, or were crushed between the elevator car, hoistway (shaft), and counterweights. Door restrictors prevent surfers and do-it-yourself escapees from forcing open the doors of an elevator car when it is between floors.

Door restrictors can be opened from outside of the car doors or on top of the elevator by rescue personnel familiar with their operation. Hydraulic tools or other types of rescue tools that apply force should not be used because they can cause unnecessary damage. Door restrictors can be operated by thoroughly trained firefighters familiar with the risks of removing occupants of stalled elevators and the dangers of using elevators during fires. If an elevator mechanic is not available in a timely period and rescuers decide to remove occupants, great care should be taken.

Types of Door Restrictors

There are two basic types of door restrictors, the clutch type and the angle iron restrictor (commonly found on Otis elevators). Clutch-type restrictors consist of a latch located at the top of the car doors (figure 7.36). The key to releasing this type of restrictor is to duplicate the pressure of the hoistway door release rollers on the restrictor release located on the car door. When the release is within reach, simply depress it and push open the car doors. If the elevator is too far above or below the floor landing to reach the release by hand, a pike pole can be used to depress the release. The second type of restrictor consists of projections fastened to the car door and hoistway similar to an angle iron. When an elevator equipped with this type of restrictor is outside its landing zone, the car door restricting angle will strike the angle iron on the shaftway doors and the shaft, preventing the car door from opening more than four inches. On older installations, the restrictor cannot be manually released.

FIGURE 7.36 Clutch-type restrictors consist of a latch located at the top of the car doors.

Some Otis elevators, however, have restrictors that are hinged and spring-loaded. Depressing the angle on the car door will allow it to clear the angle iron fastened to the hoistway and hoistway doors. There are other types of door restrictors on the market including an electromechanical type. Fire departments should contact elevator mechanics in their area to become familiar with the types of door restrictors that they may encounter. They should also identify elevator mechanics and their emergency contact information for assistance in elevator emergencies. Information about elevator operations and door restrictors in particular can be found in ASME Standard A17.1.

Door Restrictor Controversy

While door restrictors have helped eliminate one hazard associated with elevators, there is great controversy about their overall safety during stuck elevator or other emergency rescue operations. The same door restrictors that prevent occupants from exiting the elevator between floors in some cases can also prevent firefighters from opening elevator doors to rescue occupants if the elevator is between floors. Certain types of door restrictors are better than others, but consideration needs to be given to their benefits versus risks to occupants when an elevator is stuck and requires firefighter rescue.

Rescue Team Organization and Training

Firefighter personnel who are called upon to respond to elevator emergencies should have received training according to ASME A17.4. In some jurisdictions, elevator rescues are the responsibility of the fire department's rescue squad. In Baltimore City, Rescue 1 responds to reports of persons trapped in elevators. The following is a list of basic equipment recommended by ASME A17.4 for rescue teams responding to elevator emergencies.

- Short extension ladder
- Collapsible or folding ladder
- Hoistway door unlocking devices (elevator door interlock release keys)
- Two-way radios
- Safety belts
- Lifelines
- Forcible entry tools
- Flashlights

Rescue teams should be provided training in the proper procedures for evacuating persons from elevators and prepare them for actual emergency situations. Those instructors with experience and expertise in elevator operation or rescue should be utilized, including elevator maintenance or service personnel. Once training is completed, drills should be held simulating a variety of conditions in which a rescue may need to be performed. These drills should be conducted to determine the effectiveness of the rescue operations and the organization.

EVACUATION PROCEDURES

Each year, nearly ten thousand people in the United States end up in emergency rooms as a result of injuries due to elevator- or escalator-related accidents. Encounters include tripping, caught clothing, being hit by closing doors, or falling down an elevator shaft when trying to exit a stalled elevator car. Elevator mechanics, inspectors, and firefighters can also be among the injured if they do not operate safely around elevator equipment. The odds of someone in the general public getting stuck in a stalled elevator are about once in a lifetime.

The first thing to be done after arriving on scene is to establish communications with those people "trapped" in the elevator car(s). I can personally attest to this as an important function of the rescue team. Several years ago, I was in an elevator in a parking garage at Baltimore Washington International Airport (BWI) with my wife, daughter, and mother. The elevator stopped and the doors would not open. We used the emergency phone and reached a recording, so we never got to talk to an actual person to know if anyone received our call for assistance. We continued to ring the elevator emergency bell but did not hear from anyone for over thirty minutes. Then, all of a sudden, firefighters pried open the doors and we were rescued (I am hoping that was my once-in-a-lifetime entrapment). It would have been so much more comfortable had someone been in contact with us to let us know what was happening. Communications with the people in the elevator car(s) is two-way. You are trying to reassure them that help is on the way and you need some information from them. First of all, the things you should tell the people "trapped" are listed below:

- They are not in any danger.
- Rescue personnel are working to free them from the car.

- They should stay away from doors as they may be opened.
- They should not smoke in the elevator car.

Information that is sought from the occupants of the car includes:

- How many people are in the car?
- Are any of the occupants sick or injured?
- Are lights working in the car?
- If known, what is the position of the elevator in the hoistway?

During the rescue operations, occupants should be continually informed of the progress and reassured of their safety.

The preferred safe practice for evacuating people from stalled elevators is to bring the car to a landing. This can be accomplished using either the firefighter service function of the elevator or inspection controls. No other means should be used unless under the direct supervision of experienced elevator personnel. Before any of the above-mentioned operations are undertaken, the mainline disconnect switch should be placed in the "On" position (closed). If the elevator car is equipped with an emergency "Stop" switch, it should be placed in the "Run" position. If the elevator has fire service, the first action would be to activate Phase I recall from the main elevator lobby in an attempt to bring the car to the lobby.

POWER FAILURE

If an elevator car has stalled because of a power failure in the building and the building has an emergency generator, it may be possible to run elevators by turning on the recall switch in the elevator lobby on the ground floor. If the elevators have emergency power and there is a manual selection switch, you may be able to bring the elevators to the ground floor one by one.

If none of the above-mentioned actions work to free persons from the elevator, then firefighters will need to wait for trained elevator mechanics to arrive or notify a trained elevator rescue team from the facility or the fire department. Firefighters can be reasonably safe during elevator operations if they follow the following safety rules:

- Never take an elevator to the fire floor.
- Never pass the fire floor in an elevator.
- Never return in an up elevator except on independent control.
- Put every elevator in firefighter service control.
- Always activate the emergency stop before escaping a stalled elevator.
- Never overcrowd an elevator or exceed its operational weight.

8 Standby Power and Emergency Lighting

INTRODUCTION TO STANDBY POWER AND EMERGENCY LIGHTING

Building codes, including the *International Building Code* (IBC) and NFPA 5000; fire codes, including the *International Fire Code* (IFC), NFPA 1, and the NFPA *Life Safety Code*, require that emergency lighting be provided for the means of egress in almost all types of occupancies. Emergency lighting can be provided with the use of battery pack units, emergency generators, combinations, and, in some cases, two separated dependable public power sources. Emergency lighting can be either in continuous operation or capable of repeated automatic operation without needing to be turned on manually. Where battery-powered exit lights or battery-operated emergency lighting units are used, they are required to be listed to the requirements of Underwriters Laboratory (UL) 924, *Standard for Emergency Lighting and Power Equipment*. Other standards that apply to emergency lighting, generators, battery-powered units, and fuel sources for generators are found in the following codes. NFPA 110 is the Standard for *Emergency and Standby Power Systems* and covers generators. NFPA 111 is the Standard on *Stored Electrical Energy Emergency and Standby Power Systems* and covers battery-operated emergency lighting packs. NFPA 30, *Flammable and Combustible Liquids Code*, is consulted when generators are used that are fueled with liquid petroleum fuels. NFPA 37, *Standard for Installation and Use of Stationary Combustion Engines and Gas Turbines*, is also used when generators are provided for emergency power. When generators are fueled by natural gas, NFPA 54, the *National Fuel Gas Code*, applies. If the fuel is liquefied petroleum gas, then NFPA 58, *Liquefied Petroleum Gas Code*, is used. When emergency lighting systems are located in health care facilities, NFPA 99, the *Standard for Health Care Facilities*, requirements apply. The *International Building Code* (IBC) has requirements for emergency lighting for the means of egress. Because all of the emergency lighting systems involve the use of electricity in one way or another, NFPA 70, the *National Electrical Code*, would apply to installations as well. Firefighters responding to alarms and other emergencies should take note of or ask about emergency lighting sources during visits to buildings and, if time permits, make sure that those systems in place are adequate or are being maintained. Spot checks can be conducted on battery-powered units using the test button to see if they are working. Getting occupants out of a building when an emergency occurs is vitally important, and the egress system should be illuminated by emergency lighting when the primary power fails.

CODE REQUIREMENTS

NFPA 101, the *Life Safety Code*, is the primary source of code requirements for the installation of emergency lighting in various occupancies. Emergency lighting is required where individual occupancy chapters require them and in underground or windowless buildings, high-rise buildings, at doors equipped with delayed egress locks, and in smoke-proof enclosures. Almost all occupancy chapters have some emergency lighting requirements. Most chapters that do have emergency lighting requirements refer to chapter 7, "Means of Egress," section 7.9, for details. NFPA 70 lists five types of emergency power, storage batteries, generator set(s), uninterruptible power supplies, a separate service, and unit equipment. NFPA 101, the *Life Safety Code*, does not recognize a separate service as an acceptable emergency power source for those occupancies covered by NFPA 101. The *International Building Code* (IBC) and *Life Safety Code* require that emergency lighting provide initial illumination in all parts of the egress system, including exit access corridors, stairwells, stairwell access doors, passageways, at delayed egress doors, aisles in rooms, and in rooms or spaces that require two or more means of egress. The portion of the exterior exit discharge immediately adjacent to exit discharge doorways should also have emergency lighting. Egress components along the egress path are required to have lighting density at least an average of 1 foot candle (10.8 lux) and a minimum at any point of 0.1 foot candle (1.1 lux) measured along the path of egress at the floor level. Illumination levels are not allowed to be reduced to less than an average of 0.6 foot candle (6.5 lux) and at any point not less than 0.06 foot candle (0.65 lux).

Codes require that emergency power sources be available for certain occupancies, such as hospitals and nursing homes, in the event that normal power is lost. Exit signs are also required to be provided with emergency lighting in occupancies where emergency lighting is required for the egress system. This can be accomplished by emergency lighting units with their own battery backup, emergency lighting battery packs providing light for the exit signs, photoluminescent exit signs, or placing signs on the emergency circuits for the emergency generator. Emergency lighting is required to provide illumination for a period of not less than ninety minutes. Required illumination for the emergency egress system will be automatic in the event that normal power is lost, whether public power is lost or building breakers are shut off. Where emergency lighting depends on the change from one energy source to another, such as public utility to emergency generator, this change over must take place in ten seconds or less.

GENERATORS

When present, generators provide the best opportunity for emergency lighting and other emergency power requirements in occupancies (figure 8.1). Generators are usually engine driven and are part of a system that includes a prime mover (driver), a governor, a generator attached, transfer equipment, and a distribution system and protective equipment. Generally, drivers for generators are powered by gasoline, natural gas, or diesel fuel. Engines powered by gasoline and natural gas are less costly than diesel engines. However, they have distinct disadvantages. For example,

FIGURE 8.1 When present, generators provide the best opportunity for emergency lighting and other emergency power requirements in occupancies.

gasoline is highly flammable and more likely to be involved in a fire than diesel fuel. Natural gas is generally not stored on-site, so it is quite possible that the supply could be interrupted by the same emergency that caused the loss of normal power to the facility. When generators are located inside a room that a person can walk into, battery-powered emergency lighting units are required to be provided. Governors help maintain a constant speed of the driver during the entire power outage, varying the fuel input into the driver. Transfer equipment is required to sense the loss of normal power and signal the driver of the generator to start. The switch transfers the electrical load in the building to the emergency power grid. When normal power is restored to the building, the transfer switch once again transfers power back to the normal power grid. Generators are utilized when emergency power is required for long periods of time. When a generator is used in a building for emergency power, it must also supply power for life safety systems, including fire alarm, exit lights, emergency lighting, fire pump (if present), elevators, public address systems, critical heating, smoke evacuation systems, and stair pressurization. Wiring for emergency lighting must be independent of all other wiring and needs to be installed in separate raceways, boxes, and cabinets except at transfer switches and exit lighting fixtures supplied from normal and emergency power sources. Generators used for life safety systems must be automatic, starting and coming on-line within ten seconds of the loss of the public utility.

Fuel tank requirements for emergency generators will be found in NFPA 30 and NFPA 37. There are requirements for remote fuel shutoffs for fuel supplies. Restrictions exist as to size of tanks attached to engines inside of buildings and on roofs of buildings. Quantities can be in the thousands of gallons, requiring diking and leak detection alarms among the safety devices installed (figure 8.2). Firefighters should be aware that fuel supplies for generators can result in hazardous materials

FIGURE 8.2 Generator fuel tank quantities can be in the thousands of gallons, requiring diking and leak detection alarms among the safety devices installed.

concerns within structures that use them. Fuel tank rooms, depending on the size of the tanks inside, may require rated enclosures ranging from one hour to four hours. They should also be equipped with suppression systems. Annunciator panels showing operating conditions of generators are located in the generator room and in the fire command center of high-rise buildings. Preplanning of buildings is the most effective way to learn the locations of generators and fuel supplies. This can also be determined during plans reviews for new construction and remodeling of structures. However, responders can also make mental or written notes of the fire protection and life safety features of buildings as they respond to facilities and buildings for emergency calls.

BATTERY PACKS

In the absence of a generator in a building, emergency lighting, where required, is usually provided by battery-powered lighting units located throughout the egress system of an occupancy (figure 8.3). These lighting units are provided for the evacuation of occupants and are really not intended to provide light for emergency responders. Battery-powered lighting units are hooked into the building's electrical supply system and come on upon loss of power to the circuit they are hooked to. If working properly, activation is almost instantaneous. Light is required for a duration of a minimum of ninety minutes. These lighting units are equipped with battery chargers that automatically charge up the batteries when public power is restored to the building. There are two basic types of rechargeable batteries used in battery-powered emergency lighting units, lead-acid or nickel-cadmium. Lead-acid batteries are cheaper and take up less space but do not last as long, are not as durable, and require more maintenance than nickel-cadmium. Battery-powered emergency lighting units need to be tested and maintained periodically to ensure that they are

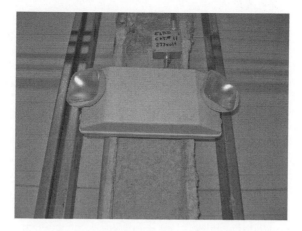

FIGURE 8.3 In the absence of a generator in a building, emergency lighting, where required, is usually provided by battery-powered lighting units located throughout the egress system of an occupancy.

FIGURE 8.4 NFPA 101 and the *International Building Code* require that a remote emergency and standby power status indicator be provided in the fire command center for all generators serving a particular building or complex.

ready to operate when needed. Batteries used in battery-powered emergency lighting units are required to be reliable types of rechargeable batteries with proper charging equipment. Batteries are required to be listed for emergency lighting purposes and comply with NFPA 70. NFPA 1 allows for emergency lighting to be fixed in place or portable and allows flashlights for portable emergency lighting in outdoor fire pump units.

FIRE COMMAND CENTERS

NFPA 101 and the *International Building Code* require that a remote emergency and standby power status indicator be provided in the fire command center for all generators serving a particular building or complex (figure 8.4). This status indicator will have an audible alarm and indicator lights associated with the device. Status indicators should be provided for high coolant temperature, low coolant temperature, low oil pressure, over crank, over speed, ground fault, circuit breaker open, engine running, low fuel day tank, rupture basin leak, battery charger AC failure, emergency stop, and generator online.

9 HVAC, Smoke Control, and Other Building Systems

INTRODUCTION

Design features of buildings are important factors in the movement of fire and smoke within a building and the ability of firefighters to conduct suppression operations and remove smoke effectively. Building obstacles that make ventilation difficult allow smoke to accumulate and obscure firefighters' vision. Windowless buildings and basements present difficult circumstances for firefighting operations and make smoke removal difficult as well (figure 9.1). Designers should take design issues into consideration and take actions to mitigate problem areas during the design phase of a building project. Building designs that lack natural ventilation or ventilation built into the building HVAC system allow for the accumulation of dense smoke and intense heat, which hamper firefighting operations and endanger building occupants who may become trapped in the building. One of the most important firefighting tactics for suppression of a fire is the venting of smoke and hot gases from a building so that suppression operations can be accomplished efficiently and safely. Building designers need to be conscious of important ventilation functions and design an effective means of venting smoke and hot gases from a building through the use of the building HVAC system or a dedicated smoke exhaust system. NFPA 204, *Standard for Smoke and Heat Venting*, should be consulted when designing buildings. Once a building is completed, building owners or managers should see that firefighters are trained in the use of building HVAC, smoke venting, control equipment, and operational controls.

No matter what type of occupancy you may be responding to, you may be exposed to the potential hazards associated with HVAC equipment. HVAC requirements in the codes are designed to accomplish the following:

- Keep smoke from being spread through air ducts inside a building or from the outside in.
- Insure the integrity of building fire protection construction such as floors, partitions, roofs, walls, and floor or roof-ceiling assemblies that may be affected by air duct systems.
- Reduce the ignition sources and combustible materials within the air duct systems.
- Use of the air duct system in a building for smoke control.

FIGURE 9.1 Building obstacles that make ventilation difficult allow smoke to accumulate and obscure firefighters' vision. Windowless buildings and basements present difficult circumstances for firefighting operations.

Firefighters should become familiar with the operations of HVAC and smoke control equipment and any associated dangers to firefighting operations (figure 9.2). They should also understand how these systems may help them in the control of air flow in the building or removal of smoke from a building. HVAC and smoke control systems do not move air as fast as smoke ejectors or positive-pressure ventilation fans. Air movement systems within a building have the potential to move smoke, hot gases, and flame from the point of origin to unaffected parts of a building

FIGURE 9.2 Firefighters should become familiar with the operations HVAC and smoke control equipment and any associated dangers to firefighting operations.

and to supply fresh air to the fire area, which may support continued combustion. This is why HVAC systems are equipped with safety devices and firefighter controls designed to prevent this from happening. HVAC systems and fire and smoke dampers are located behind the scenes in mechanical areas of the building and not readily visible or always accessible to firefighters. HVAC systems are usually located on the roof of a building, in a penthouse, in the basement, or in some cases on individual floors in mechanical spaces. Firefighter fan controls may be provided at the fire alarm annunciator, fire alarm panel, or in the fire command center in high-rise buildings. If smoke gets into ducts of the HVAC system, the fans may shut down automatically before firefighters arrive. HVAC systems are equipped with duct detectors designed to detect smoke in the ducts and shut down HVAC fans, automatically preventing the spread of smoke and hot fire gases through the duct work (figure 9.3). If fans have not shut down when firefighters arrive, they can activate control switches that will shut down the HVAC systems (figure 9.4). Firefighters may also have the ability to place the HVAC systems back in the automatic mode. Switches are not normally provided to firefighters for turning the systems back on directly. This is because of the need to have both intake air and exhaust air balanced to avoid damage to building components and systems. It is desirable that building maintenance personnel be present when HVAC systems are restarted manually after being shut down.

Building HVAC systems can be used to exhaust smoke where designed to do so. Buildings in which 100 percent outside air is provided, such as laboratory buildings, have the capability of removing smoke automatically if the proper doors are open. At a recent high-rise fire in a major university research building, corridors were filled with smoke and there were no windows that opened in the building to vent the smoke. Firefighters considered breaking out windows to exhaust the smoke from the building.

FIGURE 9.3 HVAC systems are equipped with duct detectors designed to detect smoke in the ducts and shut down HVAC fans automatically, preventing the spread of smoke and hot fire gases through the ductwork.

FIGURE 9.4 If fans have not shut down when firefighters arrive, they may have access to control switches that would override the building system controls and shut down the HVAC systems.

However, all of the labs in the research building have 100 percent outside air (air is not recirculated) and air is replaced every ten minutes with fresh outside air in the lab spaces. It was suggested by building management that doors to labs be opened and lab exhaust be allowed to remove the smoke from the corridors. This process worked very well and in a short period of time the corridors were clear of smoke. Information about HVAC capabilities or the presence of lab exhaust may not be readily available to responding firefighters and requests for assistance from building maintenance personnel should be made early in an incident. Building management can also provide written instructions on HVAC operation and locate the information near the keys for the HVAC shutdown switches or fire command centers in high-rise buildings. Plans reviewers should coordinate locations and types of switches that should be made available to responding firefighters during the plans review stage of the project. It would also be helpful if switch types and locations were standardized within a community to make training of firefighters in the use of the systems just that much easier.

Other issues to watch for during plans review include location of the fresh air intake(s), inspection access panels, smoke dampers, fire dampers, and plenums. Fresh air intakes should be located so that smoke or hot gases from an adjoining building or other fire will not be drawn into the building. Further considerations of air intake location involve potential acts of terrorism. Air intakes should be located up high on the building so that a terrorist is not able to easily reach them to induce a chemical, biological, or radiological material into the building's HVAC system. Locating intakes high also reduces the chances of unwanted smells and odors, such as vehicle exhaust from the street, getting into the building and causing odor complaints from occupants. We have one such problem location at the University of Maryland, Baltimore. Several tall buildings located in a "U"-shaped configuration by the layout of the streets create a "canyon" effect. There is usually little air movement in the area. This also happens to

FIGURE 9.5 Trucks often sit idling, waiting to unload, and the exhaust fumes are sucked into a building air-handling unit that has an intake just one story above the street.

be in an area located adjacent to a busy loading dock. Trucks often sit idling, waiting to unload, and the exhaust fumes are sucked into a building air-handling unit that has an intake just one story above the street (figure 9.5). Environmental Health & Safety at the university receives numerous complaints of exhaust and other odors in the building on a regular basis. Designers should have recognized and addressed this potential problem during the design of the building projects in the area. Part of the design process of a building is to determine how the other buildings in the area and other environmental factors will affect the building under design. Consideration also needs to be given to the affects the building under design may have on other buildings in the area. Firefighters should be careful when responding to alarms in existing buildings where building air exhaust panels are located at or near the ground level. Toxic materials, terrorist agents, and smoke can be exhausted out these locations and be a hazard to firefighters who are not wearing proper respiratory protection.

Smoke and fire dampers may be installed in HVAC systems at rated fire and smoke partitions to control the movement of smoke and fire gases through a building. Generally, fire dampers are required for fire walls and smoke dampers are required for smoke separations and barriers. Fire dampers are equipped with fusible links that melt from the heat of a fire and close the damper (figure 9.6). Smoke dampers are hooked into the building fire alarm system and close automatically with the activation of the building fire alarm or smoke detectors located on either side of a smoke partition (figure 9.7). Each fire damper, smoke damper, and smoke detector needs to have service openings provided large enough to allow for maintenance and resetting of the devices. If service openings are in rated walls or ceilings, then access panels need to be rated for the ceiling or wall they are placed in. Smoke dampers are required in various locations according to NFPA fire and building codes. NFPA 90A requires smoke dampers to be installed in systems having a capacity of 15,000 ft^3/min (7080 L/s) or greater where ducts penetrate smoke barriers.

FIGURE 9.6 Fire dampers are equipped with fusible links that melt from the heat of a fire and close the damper.

Preplanning buildings within first due areas can provide an opportunity for fire-fighters to record information about the HVAC and other systems within a target building. While this type of information is mundane and often overlooked, it can be of great benefit for restricting smoke movement within a building or exhausting smoke from a building during a fire. NFPA 90A is the standard for the *Installation of Air-Conditioning and Ventilating Systems* and generally covers buildings over 25,000 square feet. Buildings of Type III, IV, and V construction, regardless of volume, are also covered by NFPA 90A. Type III construction includes buildings for which exterior walls and structural elements that are portions of exterior walls are

FIGURE 9.7 Smoke dampers are hooked into the building fire alarm system and close automatically with the activation of the building fire alarm or smoke detectors located on both sides of a smoke partition.

of approved noncombustible or limited-combustible materials. Interior construction may be entirely or partially of wood or other combustible material or noncombustible or limited combustible materials. Type IV construction consists of fire walls, exterior walls, and interior bearing walls, and structural elements are comprised of approved noncombustible or limited-combustible materials. Type V construction consists of materials in which structural elements are entirely or partially of wood or other approved material. NFPA 90B is the standard for *Installation of Warm Air Heating and Air-Conditioning Systems* and applies to one- and two-family dwellings and buildings over 25,000 square feet. Building codes also contain information about installation requirements for HVAC systems.

DUCT DETECTORS

Smoke within an air duct system can not only be deadly to building occupants but also to firefighters and other responders. Smoke detectors specifically designed for use in a duct system are strategically installed within the ductwork to detect smoke in the HVAC system. Generally, duct detectors are installed in the main supply ducts, downstream of air filters or air cleaners, and in the main return ducts before the point where the air is exhausted from the building. When smoke is detected, the fan(s) associated with the air movement within the duct system are shut down. Detectors may be designed or programmed to activate the building fire alarm system when smoke is detected in the duct system or the *National Fire Alarm Code* allows for a supervisory signal to be sent when the smoke detector is activated and the fan is shut down. Since the HVAC system is a closed system and things other than smoke can trigger smoke detectors, using the supervisory signal will help to reduce the chances of false alarms. Because of the dilution effect of the volume of air in the duct system, the detectors are not a very effective early warning system anyway. If there really is a fire within the building, other detection devices or sprinkler systems will activate the fire alarm system in the building or someone will see the smoke or fire in the building and activate a manual pull station. If the building does not have a sprinkler system or other detection devices, then it would be wise to have the duct detectors activate the building fire alarm system instead of the supervisory signal. Buildings where people live and sleep should also have duct detectors that activate the building fire alarm system. NFPA 72, the *National Fire Alarm Code*, and NFPA 70, the *National Electrical Code*, should be consulted for guidance in placement of smoke detectors and wiring of HVAC and other systems within a building.

SMOKE DAMPERS

According to the IBC, smoke dampers are installed to isolate the HVAC equipment from the rest of the system to restrict the movement of smoke. Smoke dampers have a listed smoke detector installed in the duct within five feet of the damper, and they are hooked into the building fire alarm system. When dampers are installed above doors in smoke barriers, the damper will close upon activation of smoke detectors on either side of the barrier. If a damper is installed in an unducted opening, a smoke detector needs to be installed within five feet of the damper. If a complete smoke detection system is installed, it is permitted to cause the dampers to close upon activation of any smoke detector in the system. When ductwork within a building passes through a smoke barrier, smoke dampers should be installed to shut that damper if smoke is present. Smoke dampers are electronic devices connected to the building fire alarm system and close automatically when the smoke detector connected to them activates. Smoke dampers are required by NFPA 90A, *Standard for the Installation of Air-Conditioning and Ventilating Systems*. Smoke dampers should be installed in systems that have an air-moving capacity of greater than 15,000 ft³/min (7080 L/s) to isolate air-moving equipment from the rest of the system and restrict the circulation of smoke and fire gases.

Building codes also have requirements for smoke dampers and care should be exercised when searching the building code, NFPA, and other codes so that all of the requirements and exceptions are noted. Smoke dampers are placed in required smoke partitions where ductwork passes through. In the past, smoke dampers have been placed in areas where they were not required because the project engineers and plans reviewers did not look at exceptions to code requirements. The *International Building Code* requires smoke dampers in storage rooms where ducts pass through the rated walls of the room. A building under construction had seven smoke dampers installed in a storage room that was eight by ten feet. There was an exception to the smoke damper requirements for storage rooms located on the next page of the code. In this case, since the storage areas were less than 10 percent of the floor area, they were not required to have smoke dampers. While it does not hurt to go beyond code requirements, because codes are just minimum standards, in some cases the money saved could be better spent in other locations of the project. Additional fire alarm devices that are not required provide no increased life safety to building occupants and can also cause additional maintenance costs. Keep in mind that there are often conflicts between fire and building codes. That is why it is important to thoroughly research the code books before making decisions during plans review. Generally, when there is a conflict between codes, the more stringent code requirement takes precedence. However, the authority having jurisdiction (AHJ) may waive a code requirement with justification. The fact that one code cites a lesser requirement could provide justification.

FIRE DAMPERS

Fire dampers are mechanical devices operated with a fusible link and placed in ducts where the ductwork passes through fire walls with a two-hour rating or greater (figure 9.8). One-hour rated fire walls do not require fire dampers. They are also used in openings in two-hour or greater rated fire walls without ducts. Underwriters Laboratory (UL) Pamphlet 555, *Standard for Safety Fire Dampers*, contains the rating procedures for fire dampers. Fire dampers, as with other fire protection and alarm systems used in a building, must be listed by a recognized testing laboratory and labeled for their intended use. The *International Building Code* requires that dampers be rated at 1.5 hours where the fire wall is three hours or less. Where the fire wall rating is greater than three hours, the damper should be rated at three hours. Fire dampers are usually not connected to the fire alarm system and have to

FIGURE 9.8 Fire dampers are mechanical devices operated with a fusible link and placed in ducts where the ductwork passes through fire walls with a two-hour rating or greater. Access panels need to be provided to reset dampers when they are activated.

be manually reset and a new fused link attached once they operate. For this reason, it is not only important that the dampers be tested following new installation, an access panel needs to be provided to allow for maintenance and resetting of the dampers when they activate. Manufacturer's instructions and the UL listings should be followed closely when fire dampers are installed to make sure the listing is not violated and the damper will work as intended.

SMOKE CONTROL SYSTEMS

Smoke control systems (sometimes referred to as "smoke management systems") are mechanical systems that control the movement of smoke during a fire. Most are intended to protect occupants while they are evacuating or being sheltered in place. The most common systems referenced in current codes are atrium smoke exhaust systems and stair-pressurization systems. In specialized cases, zoned smoke control systems may be provided in a building as well. These systems feature zones or floors that are either pressurized or exhausted to keep smoke from spreading. Smoke control can also be accomplished by manual means. For example, a large shopping center in Maryland has so much space volume above the ceilings and above the corridors in the mall that smoke does not reach the exit level height for quite some time. In fact, they do not evacuate the entire mall during a fire alarm; the building actually has fire alarm zones, and only a portion of the building is evacuated at one time. Actual smoke testing was conducted to prove that the system worked. During the testing, smoke never reached the level of the exits, which allowed people to exit the building safely before their exit path was blocked by smoke. The IBC contains mandatory provisions for smoke control systems.

Designers can find NFPA's detailed provisions in two mandatory documents, the *Recommended Practice for Smoke Control Systems* (NFPA 92A) and the *Guide for Smoke Management Systems in Malls, Atria, and Large Areas* (NFPA 92B). The manual controls required or provided for smoke control systems are a primary consideration for the fire service. These manual controls can override automatic controls that activate these systems. When fire department personnel arrive, they can assess whether the automatic modes are functioning as intended. Incident commanders may then use the manual controls to select a different mode or turn any given zone off. It is imperative that these controls override any manual or automatic controls at other locations in the building. A simple, straightforward control panel with manual switches for the smoke control system(s) will assist a firefighter who may be trying to decipher how the controls work after just awakening in the middle of the night (figure 9.9). Similar to annunciators, the fire department may have specific requirements or recommendations and may prefer uniformity of smoke control and HVAC panels within their jurisdiction.

Both the IBC and NFPA 92A call for status indicators for each fan, smoke damper, and other smoke control device in the building. The IBC requires individual controls for each of these devices but permits them to be combined for complex systems (figure 9.10). A system need not be very large to be considered complex. A good simple panel layout might feature a single switch for each system or zone. Each different position of the switch places the system in a given mode and the corresponding activation or setting of the individual devices would be configured "behind the scenes." For example, a stair-pressurization system might contain a three-position switch for each of three modes: "automatic," "pressurize," and "off." Zoned smoke control systems are often arranged with each floor as a separate zone. In other cases where there are smoke barriers, a floor may be split into multiple zones. All zones

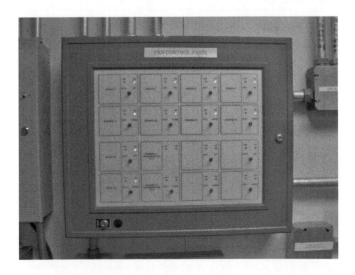

FIGURE 9.9 A simple, straightforward control panel with manual switches for the smoke control system(s) will assist a firefighter who may be trying to decipher how the controls work after just awakening in the middle of the night.

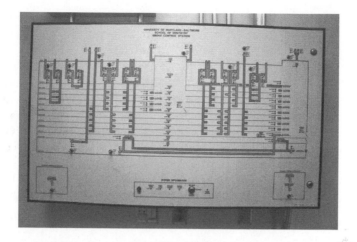

FIGURE 9.10 The IBC requires individual controls for each of these smoke control devices but permits them to be combined for complex systems.

should be indicated on a graphic display, either on or adjacent to the smoke control panel. Designers should not confuse smoke control systems with smoke or heat venting systems. The latter are mechanical systems for the removal of smoke. They are often arranged to activate only manually. In some cases, they only remove smoke after a fire has been extinguished.

"Dilution" is a term sometimes used for smoke purging, smoke removal, smoke exhaust, smoke extraction, or smoke venting. Dilution can be beneficial to the fire service for removing smoke after a fire has been extinguished. HVAC systems that are only designed for daily air management in a building will not be much help in removing smoke from the building unless they provide 100 percent outside air and have multiple air exchanges per hour. Building fires just produce too much smoke for the HVAC system to remove in a timely fashion. Engineered smoke removal systems need to be designed for a particular building if the intention is to remove smoke during and following extinguishment of a building fire. It is important to remove smoke quickly after a fire is extinguished so that firefighters can investigate to make sure the fire is completely extinguished. Once smoke removal systems are installed, they need to be tested to make sure they will function as designed and as desired by responding firefighters. Smoke should be used to provide a realistic test.

STAIR PRESSURIZATION

High-rise buildings require that emergency exit stairwells be smoke-proof. One method of accomplishing this is by pressurizing the stairwell to keep most of the smoke and products of combustion out of the stairs during a fire. As people enter stairwells to evacuate from a fire, the smoke in the building generated by the fire can follow them into the stairwell. If doors to stairwells are left open, smoke can quickly migrate into stairwells, making them impassable. It is the goal of the design of the stair-pressurization system for a building to maintain the stairwells in a tenable state in the event of a building fire until people have safely evacuated. Pressurized stairwells are designed to limit the amount of smoke entering the stairwell but not keep the stairwell totally smoke free. According to NFPA 101, the *Life Safety Code*, stair-pressurization systems are required to be "engineered approved systems with a design pressure difference across the barrier of not less than 0.05 in. water column (12.5 N/m^2) in sprinkled buildings and 0.10 in water column (25. N/m^2) in non-sprinkled buildings." Systems are required to be further engineered to allow for the proper pressure differential across the barrier but allowing for the stairwell door to be opened with a force of 30 lbf (133 N) or less, which is part of the acceptance testing of stair pressurization. This is to make sure that occupants can easily open doors during evacuations. This test can be accomplished with the use of a fish scale. From the inside of the stairwell, place the hook end of the scale on the door handle. Pull the handle to obtain a reading of the force required to open the door (figure 9.11).

Since pressurization fans and ductwork are life safety systems, they must be installed to protect survival during a fire. They may be installed outside the building, with ductwork connecting them directly to the stairwell and enclosed in noncombustible construction. Fans may be installed inside the stairwell enclosure with intake and exhaust air vented directly outside the building or through ductwork enclosed in two-hour rated construction. Equipment and ductwork may also be installed within the building provided that certain conditions are met. Pressurization equipment and ductwork are installed within a two-hour rated enclosure. If the building is fully sprinkled, the rated enclosure requirement is reduced to one hour. Even if pressurization equipment and ductwork is installed in a mechanical space, they still need to be separated from that space by the properly rated one- or two-hour enclosure. Stair-pressurization fans need to be supplied with emergency power in case the building power is turned off or lost. Stair pressurization is activated by

FIGURE 9.11 Door force to open tests can be conducted with the use of a fish scale.

several means. One method is by a smoke detector hooked into the building fire alarm system installed within ten feet of the stairwell opening. Manual controls also need to be provided in the fire command center for firefighter use to manually start the stair pressurization if needed (figure 9.4). Stair pressurization can also be started by activation of water flow devices within the building or by a general fire evacuation alarm within the building. It is my recommendation that the stair pressurization be activated by the building fire alarm system any time it is activated.

CHLORINATED FLUOROCARBONS (CFC)

Chlorinated fluorocarbons are chemical refrigerants suspected of depleting the Earth's ozone layer. Chlorofluorocarbons (CFCs) were developed in 1928 as a replacement for ammonia (NH_3), chloromethane (CH_3Cl), and sulfur dioxide (SO_2), which are toxic but were in common use at the time as refrigerants. Large buildings may utilize chlorinated fluorocarbons (CFCs) as part of the cooling system. Chloro-fluorocarbons are usually characterized by high vapor pressure (low boiling point) and density, low viscosity, and solubility in water. There are two primary CFC agents still in use in most countries, including those that signed the Montreal Protocol: CFC 11 and CFC 12. Chlorinated fluorocarbons are gases of low toxicity but not entirely inert. They may also displace oxygen in the air in a confined space or room and cause asphyxiation to those not wearing self-contained breathing apparatus (SCBA). SCBA need to be provided outside doors to areas where CFCs and other refrigerants are located (figure 9.12). Bradycardia is the usual response in human subjects inhaling 10 percent of CFC 11. It is suspected that bradycardia in man originates from irritation of the upper respiratory track and that cardiac effects can be initiated prior to absorption of CFC 11 in the lungs. It may also be a central nervous system depressant in high concentrations. By inhalation, large, acute doses have resulted in cardiac sensitization (arrhythmia or bronchial constriction), leading to death. When exposed to heat, CFC 11 can decompose to produce phosgene gas, which is highly toxic. Never enter a mechanical space where the CFC alarm has been activated without SCBA. Special CFC monitors are available that can be used to monitor the air when a CFC leak is suspected.

CFC 11 and 12 are not the only refrigerants in use today. There are approximately 180 refrigerants listed in *Wikipedia*, the free encyclopedia. They have varying levels of toxicity and almost all could displace oxygen in a confined area in the gaseous form. Firefighters should exercise caution when responding to emergencies in large buildings with reports of odors or chemicals because it could be a refrigerant release. Firefighters should always wear SCBA when entering a mechanical area looking for a leak. Material safety data sheets (MSDS) should be available on site covering the important chemical information about refrigerants. Firefighters could also make up response books with MSDS for the most common refrigerants in use in their jurisdiction. When searching for a leak in a building, watch for CFC and other types of refrigerant audible and visual alarms, indicating a possible leak (figure 9.14). There are also remote shutoff devices for shutting down refrigerant equipment in the event of the leak (figure 9.13). They are usually located in the vicinity of the CFC or refrigerant alarm.

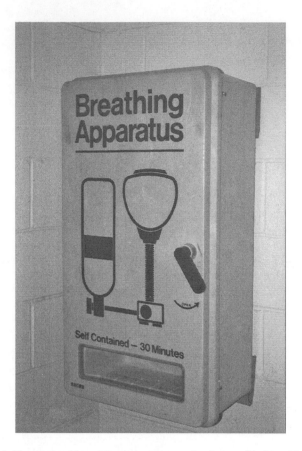

FIGURE 9.12 Self-contained breathing apparatus need to be provided outside doors to areas where CRCs and other refrigerants are located.

FIGURE 9.13 Mechanical rooms where CFCs are in use are required to have automatic leak alarms with manual shutdown options.

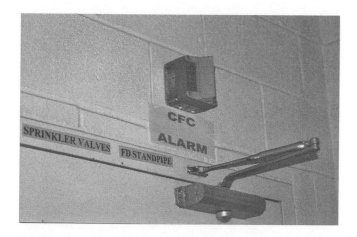

FIGURE 9.14 When searching for a leak in a building, watch for CFC and other types of refrigerant audible and visual alarms, indicating a possible leak.

NMR AND MRI RESEARCH EQUIPMENT

Research facilities in industry and at major universities in the United States and other parts of the world may conduct research utilizing magnetic resonance imaging (MRI) and nuclear magnetic resonance (NMR), which use very high-powered magnets. These magnets are much more powerful than the MRI units that most people are familiar with in visits to doctors and hospitals. MRI is a powerful diagnostic tool that uses magnetic fields and radio waves, not radiation, to create images of the body (figure 9.15). MRI also has uses outside of the medical field, such as detecting rock

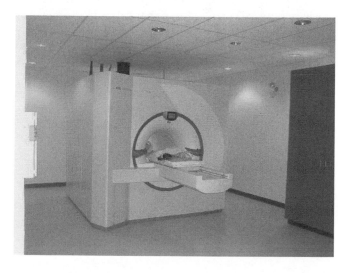

FIGURE 9.15 MRI is a powerful diagnostic tool that uses magnetic fields and radio waves, not radiation, to create images of the body.

FIGURE 9.16 The presence of these magnets and associated magnetic fields cannot be detected by the human senses.

permeability to hydrocarbons and as a nondestructive testing method to characterize the quality of products such as produce and timber.

The presence of these magnets and associated magnetic fields cannot be detected by the human senses (figure 9.16). There are no short-term health effects from the magnets or magnetic fields themselves. An important safety concern with NMR is the machine's ability to attract metal objects. The typical NMR is thirty thousand times more powerful than the Earth's gravitational field. So any metal objects in the room or brought into the room by responding firefighters can be pulled into the NMR unit (figure 9.17). Credit cards or other cards with magnetic strips on the back can be erased by exposure to the NMR or MRI. In 2001, a six-year-old boy undergoing an NMR diagnostic test died when a steel oxygen tank was suddenly pulled into the patient chamber at high speed. The oxygen tank had been brought

FIGURE 9.17 Any metal objects in the room or brought into the room by responding firefighters can be pulled into the NMR unit.

into the room during the scan and was pulled out of the doctor's hands. The young child suffered a skull fracture and brain hemorrhage and died two days later. Since then, health care workers have reported other objects being inadvertently caught in the pull of the NMR unit, such as a toolbox, vacuum cleaner, defibrillator, and wheelchair. Even small objects can be dangerous. Experts say a metal wrench pulled from just two feet away from an NMR unit will hit the machine at a speed of about thirty to forty miles per hour. The risks associated with NMR are not limited to metal objects in the room. Any magnetic-susceptible metal in or on the body can also be affected by the magnetic force from an NMR. Metallic objects in the body, like surgical wires, clips, and pacemakers, can be pulled into the magnetic field. Serious injury or death can occur if the implant is dislodged (as in an aneurysm clip in the brain) or malfunctions (interference with a pacemaker). Metal can also heat during an NMR scan and cause severe burns in the surrounding area. Jewelry, electrode leads, and foil-backed drug patches have been reported to cause serious burns during an MRI scan. Even metal fragments in tattoo ink or cosmetics can cause burns.

One important misunderstanding is the impression that an NMR unit is safe when not in use. However, an NMR is never turned off, except for maintenance (suddenly shutting the unit down can damage the very costly magnet). So even though there may not be anyone in the scanner, the machine still has a very strong magnetic force. Therefore, all personnel who may have access to a room with an NMR unit should be educated about potential dangers. If, for whatever reason, firefighters determine that the magnet needs to be turned off for rescue or other operational needs, the units are provided with a key-operated quench button. This button will shut down the magnet by exhausting the helium in the magnet to the outside of the building. This should be done as a last resort because release of the helium is very expensive and the magnet can also be damaged. Two asphyxiant gases are used in the operation of a magnet, nitrogen and helium. These gases are not toxic but can displace the oxygen in the air, causing anyone without SCBA to become asphyxiated. NMR rooms generally have low-oxygen sensors installed to warn of a gas discharge. NMR and MRI areas are marked with signs. Firefighters should watch for these signs in research facilities and make sure that they take proper precautions before entering any area where there is a magnetic field.

HIGH-PRESSURE STEAM SYSTEMS

Buildings in cities are often found heated by high-pressure steam. This steam is piped from the generating plant to the building needing heat through pipes under the streets. High-pressure steam has a minimum temperature of 212°F (324.32°C). When the steam enters the building, the pressure is approximately 150 to 180 psi. Through pressure reduction equipment in the room the pressure is lowered to 60 psi for use in the building (figure 9.18). Steam leaks can expose firefighters to dangers from the high pressure and the temperature of the steam. When a leak occurs in a mechanical area, firefighters should not enter the area unless it is necessary to rescue someone injured by the leak. Otherwise, firefighters should wait for the

FIGURE 9.18 When the steam enters the building, the pressure is approximately 150 to 180 psi. Through pressure-reduction equipment in the room, the pressure is lowered to 60 psi for use in the building.

workers from the steam-generating company to shut off the steam supply to the building. Steam systems have pressure relief valves that vent the steam directly outside at the roof level. When responding to a reported steam leak, you may see steam discharging from the top of the building, which is an indication that the relief valve is operating.

HEATING FUELS

Many buildings are heated with or have natural gas, propane, butane, diesel fuel, or heating oil in use within the building for heating and various other purposes. When a fire or leak occurs, firefighters need to know where the shutoff valves are located so they can cut off the supply of fuel to a building or portion of a building. They may also need to shut off the supply from a fuel storage tank during an emergency. Natural gas is usually supplied to a building through a fixed piping system. Main shutoff valves for natural gas may be located inside or outside the building depending on the part of the country you live in. If located inside, they are usually in a mechanical space in the basement or on the ground floor (figure 9.19). If outside, shutoffs may be anywhere around the perimeter of the building (figure 9.20). Natural gas shutoffs may also be found for selected areas of a building. They may control an entire floor or just a room or group of rooms where natural gas is used. An example would be research labs, where natural gas is piped into each lab with an emergency shutoff valve in the corridor outside the lab (figure 9.21). Propane, butane, fuel oil, and diesel fuel are usually supplied to a building from a storage tank in the building or outside the building. Shutoff valves are generally located at the storage tank. Firefighters should be familiar with the physical and chemical characteristics of fuels that may

FIGURE 9.19 If located inside, natural gas shutoffs are usually in a mechanical space in the basement or on the ground floor.

be found in local buildings. Natural gas, for example, is lighter than air. In a leak you would expect to find natural gas near the ceiling or upper floors of a building. Propane, butane, fuel oil, and diesel fuel vapors are heavier than air, so vapors should be expected to be found in basements and on lower floors of a building. When these fuels leak, firefighters need to know the flashpoint of the materials and control ignition sources because they are all flammable to some degree. Once again, MSDS would be very helpful in dealing with these materials. Fuel oil and diesel fuel can have varied formulations depending on the oil company. You need to know the brand to get the appropriate MSDS.

FIGURE 9.20 If outside, shut offs may be anywhere around the perimeter of the building.

FIGURE 9.21 An example would be research labs, where natural gas is piped into each lab, with an emergency shutoff valve in the corridor outside the lab.

ELECTRICAL EQUIPMENT

Large buildings or complexes require a great deal of electricity to power the buildings during daily operations (figure 9.22). Those power systems may be partially or fully backed up with emergency power, usually from a generator in the building. Firefighters should determine what, if any, emergency power is available in a building before having power cut to all or any section of the building for operational purposes. When primary power to a building is cut, generators are designed to come on automatically. Firefighters may think the power is off for safety purposes and

FIGURE 9.22 Large buildings or complexes require a great deal of electricity to power the buildings during daily operations.

then the generator comes on and resupplies the power to the building or parts of the building. This can create a hazard to firefighters. Firefighters should not be trying to disconnect power to commercial buildings without the electric company present or qualified maintenance personnel from a particular facility. Shutoff of high voltage requires specialized training and knowledge to be done safely and should not be attempted by firefighters. It is questionable for firefighters to be shutting off power to any structure, including private dwellings, without assistance from the power company. Firefighting operations should not be conducted around energized electrical equipment in a building because water conducts electricity, and water from hose lines in contact with energized electrical equipment can result in serious injury or death to firefighters. Directing hose streams into high-voltage transformers can result in an explosion. Fighting fires can also create runoff, and standing water can conduct electricity as well. Firefighters should exercise extreme caution when working in or near electrical transformer rooms, electrical equipment rooms, or other mechanical areas of a building. Building maintenance personnel should be consulted and preferably be present to provide guidance to firefighters when dealing with building systems.

Electricity is often a source of many fires that occur in residential and other types of occupancies. Extension cords and other devices that overload circuits result in a buildup of heat and, ultimately, a fire (figure 9.23). During responses to buildings and during inspections, personnel should look for obvious electrical hazards. During one recent trip through a business occupancy, I rounded up fifteen extension cords and several multiplug adapters and mitigated numerous situations where overloads of cords and bar strips were taking place. This was, by the way, a brand-new high-rise building! Refrigerators, microwaves, space heaters, coffeemakers, and other appliances were plugged into bar strips and extension cords. Some of the cords were grossly undersized. In many cases I was able to get everything plugged into an outlet by just moving the appliances or removing extension cords and simply

FIGURE 9.23 Extension cords and other devices that overload circuits result in a buildup of heat and, ultimately, a fire.

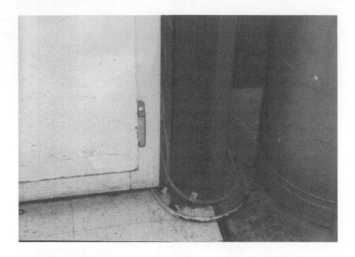

FIGURE 9.24 Extension cords should not be used to power appliances, and electric cords of any kind should not be run through doorways or ceilings.

plugging things into outlets. Appliances of any kind should be plugged directly into wall outlets. Portable lights, lamps, computer equipment, and other electronics can be plugged into bar strip–type surge protectors safely. Be on the lookout for these electrical violations and oftentimes it takes only a few minutes to correct the situation and prevent a potential fire. Extension cords and multiplug adapters are in violation of NFPA 70, the *National Electrical Code*, when used in place of permanent wiring. Extension cords are only permitted for temporary use, such as powering a computer and projector during a presentation. Extension cords should not be used to power appliances, and electric cords of any kind should not be run through doorways or ceilings (figure 9.24).

10 Inspection, Testing, and Maintenance

All fire alarm and fire protection systems, whether required or not, if installed, must be tested and maintained according to schedules provided in appropriate codes. When new systems are installed, it should not be assumed that they are installed properly for any reason. All fire alarm, fire protection, and life safety systems need to be tested to make sure they work as intended and required by code. Once systems have been accepted and occupancy permits are issued, the testing and maintenance process needs to continue according to schedules and time frames dictated by the codes to insure that systems will continue to function as intended for years to come. Owners of property are responsible for the maintenance and testing of all fire alarm, fire protection, and life safety systems. However, fire prevention personnel must provide the necessary periodic oversight to make sure that the testing and maintenance takes place.

FIRE ALARM SYSTEM MAINTENANCE AND TESTING

Fire alarm systems are tested and maintained according to NFPA 72, the *National Fire Alarm Code*. NFPA 72 outlines the types and frequency of required fire alarm tests. Newly installed or upgraded fire alarm systems need to be tested to ensure that they work properly and are installed according to code. NFPA 72 lists all tests and frequencies required for fire alarm systems, including the initial acceptance test. Installers should test the fire alarm system before the arrival of the authority having jurisdiction (AHJ) so their time will not be wasted fixing devices and components that do not work. All fire alarm devices, including pull stations, smoke detectors, heat detectors, beam detectors, and others, should be tested. Activation of the fire alarm device should register on the annunciator and fire alarm panels as well as the remote monitoring location for the fire alarm. Water flow and tamper switches in sprinkled buildings should be hooked into the fire alarm system and tested along with other devices. Smoke detectors should be tested with live smoke and heat detectors with a heat source. Duct smoke detectors, elevator lobby smoke detectors, and elevator machine room smoke and heat detectors should also be tested. All audible devices and visual devices should be tested. Audible devices should be tested with a sound meter and follow the decibel requirements found in NFPA 72. Any smoke evacuation or stair-pressurization systems present should be tested with live smoke and or pressure differential test instruments. Pressurized stair design pressure across the barrier should be not more than 0.05-in water column (12.5 Pa) in sprinkled buildings or 0.10-in water column (25 Pa) in unsprinkled buildings. These pressures should be maintainable under conditions of stack effect or wind. A test should also be conducted on stairwell and other affected doors to make sure they open within the required 30 lfb. This is important, because if the pressure in the stairwell is too great, occupants may not be able to open the doors to stairwells and could become trapped within the building.

Backup fire alarm system batteries should be given a twenty-four-hour test to determine their ability to meet code requirements if the public power service is interrupted. Oftentimes batteries are undersized because the cheapest option for battery backup is taken or calculations are not performed. During the twenty-four-hour test, the fire alarm system is taken off primary power. The event is documented on the printer at the fire alarm panel, other location, or by the AHJ. The power supply should be locked out and witnessed by the AHJ or the printer records can be viewed to insure the integrity of the test procedure. Following the twenty-four-hour period, the fire alarm system should be activated and allowed to operate for a period of five minutes. Voice alarm systems must be operated for a period of fifteen minutes. If batteries fail to support the operations of the fire alarm system for the desired period of time, they should be replaced with larger batteries. Make sure that primary power is restored to the fire alarm system following any testing procedures. The entire fire alarm system test should be documented by the installer/technician and copies made available upon demand by the AHJ.

PERIODIC TESTS

In addition to the initial acceptance testing, fire alarm systems are also required to undergo periodic testing by the fire codes. These tests should be performed by qualified fire alarm technicians. Smoke detectors should be tested for operation using smoke or an aerosol test material approved by the detector manufacturer. Sensitivity testing should be conducted to make sure each detector is within its listed and marked sensitivity range. Manufacturers' instructions should be followed when conducting sensitivity tests. Smoke detectors should always be tested in the place they were installed to ensure smoke entry into the sensing chamber and an alarm response. Specific test information can be obtained from NFPA 72, the *National Fire Alarm Code*, Chapter 10. Fire alarm system testing requirements occur monthly, quarterly, semiannually, and annually. All testing should be documented and results made available when requested by the AHJ. Samples of report forms can be found in chapter 10 of NFPA 72 and in Annex A. Persons who inspect, test, and maintain fire alarm systems are required to be qualified and experienced. Some states and local jurisdictions require that fire alarm service personnel be licensed. Fire departments should keep files or databases of all buildings where fire alarm and fire protection systems are in place. They should check at least annually to make sure all required testing is taking place in those facilities. Oftentimes fire alarm testing companies will have a vested interest in locations where they test alarm and protection systems and will tell you if someone discontinues a service contract.

SUPPRESSION SYSTEM MAINTENANCE AND TESTING

Water-based fire protection system maintenance requirements are found in NFPA 25, *Standard for Inspection, Testing, and Maintenance of Water-Based Fire Protection Systems.* Included in this standard are sprinkler systems, standpipe and hose connections, fixed water spray, and foam water. Also included in NFPA 25 are private fire service mains and appurtenances, fire pumps, water storage tanks, and valves that control system flow. NFPA 25 and NFPA 13, *Standard for the Installation of Sprinkler Systems*, are the primary sources for information about testing and maintenance of water-based fire protection systems. Not included in NFPA 25 are testing and maintenance requirements for NFPA 13D, *Standard for the Installation of Sprinkler Systems in One- and Two-Family Dwellings and Manufactured Homes*. Water spray fixed systems are also covered by NFPA 15, *Standard for Water Spray Fixed Systems for Fire Protection*. NFPA 16 is the *Standard for the Installation of Foam-Water Sprinkler and Foam-Water Spray Systems*. NFPA 17A is the *Standard for Wet Chemical Agent Extinguishing Systems*. Carbon dioxide systems are covered by NFPA 12, *Standard on Carbon Dioxide Extinguishing Systems*. NFPA 17 is the *Standard for Dry Chemical Extinguishing Systems*. Halon 1301 is covered by NFPA 12A, *Standard on Halon 1301 Fire Extinguishing Systems*. Halon 1211 is covered by NFPA 12B, *Standard on Halon 1211 Fire Extinguishing Systems*.

The importance of proper installation, acceptance testing, and a program of periodic testing and maintenance is critical to the expected functioning of all fire suppression systems, whether water based or composed of other extinguishing agents. During 1999 at the University of Maryland, Baltimore, electronics shop personnel were conducting the annual test of the fire alarm system in the law school library. A new technician pulled a manual station in a rare book room that was protected with a halon suppression system. All of the building was fully sprinkled except for the rare book room, where the halon system provided the only protection. The pull station was not marked as a halon system manual activation pull station; it looked like all of the other manual fire alarm pull stations in the building. It was, in fact, connected to the halon system, and pulling the station dumped six thousand dollars' worth of halon into the rare book room. On the surface this sounds like a mini disaster and potential loss of job for the rookie technician. But what was discovered as a result of the accidental halon discharge may have caused the discharge of the halon system to be "priceless." As it turns out, when the halon system was installed some twenty years earlier, it was not tested with an inert gas such as nitrogen to make sure it worked. Someone had left a fist-sized piece of rubber material inside the discharge piping (figure 10.1). The system had only one nozzle. When the system was activated, the pressure sent the rubber material through the discharge piping to the nozzle and obstructed the nozzle (figure 10.2). There were actually "skid" marks from the rubber material inside of the piping (figure 10.3). The halon, which was designed to discharge in less than thirty seconds, took twenty minutes to completely discharge (figure 10.4). If a fire had occurred in the rare book room, the books would have likely burned up. The moral of the story is that all suppression systems should be tested following installation and maintained according to the code requirements.

When sprinkler systems are installed new or renovations occur to existing systems, a final acceptance test needs to be conducted to insure that the sprinkler system is working as intended. Systems acceptance requirements for sprinkler systems are found in NFPA 13 chapter titled "Systems Acceptance." Hydraulic tests are required for sprinkler system piping at 200 psi (13.8 bar) for two hours. Examples of report forms are located in the same chapter of NFPA 13 as well. The AHJ does not need

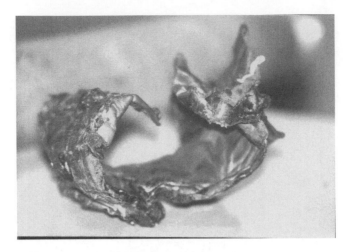

FIGURE 10.1 Someone had left a fist-sized piece of rubber material inside the discharge piping.

to be present for hydrostat testing unless they wish to. The general contractor for the project can witness the start and stop of the test and the sprinkler contractor will provide certifications based on the tests to the AHJ. If you have not received these certifications, I would ask for them during the final acceptance test. When a new sprinkler system is installed, a hydraulic design information sign should be installed on the main sprinkler system riser near the alarm check valve, dry pipe valve, pre-action valve, or deluge valve supplying the system (figure 10.5). This plate contains information about location of the design area or areas, discharge densities, required flow and residual pressure, occupancy classification, hose stream allowance, and name of installing contractor.

FIGURE 10.2 When the system was activated, the pressure sent the rubber material through the discharge piping to the nozzle and obstructed the nozzle.

FIGURE 10.3 Rubber material shown from the pipe side of the nozzle fully blocking the opening, which caused the halon to take twenty minutes to fully discharge.

Sprinkler systems require inspection of certain components, testing of others, and maintenance for specific portions on a periodic basis. Inspections may occur weekly, monthly, quarterly, annually, and at five-, ten-, and twenty-year intervals. All requirements for inspection, testing, and maintenance are found in the NFPA 25 table "Summary of Sprinkler System Inspection, Testing, and Maintenance." One of the key annual tests required of sprinkler systems is the main drain test. The main drain is usually attached to an alarm check valve. Also on the check valve are two pressure gauges. One gauge is for the supply side and one on the system side. Gauges should be attached directly to the alarm check valve in order to operate appropriately. The main drain test determines whether there has been any change

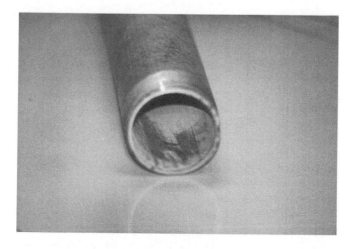

FIGURE 10.4 There were actually "skid" marks from the rubber material inside of the piping.

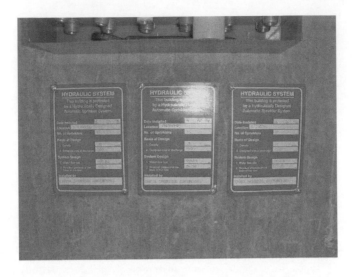

FIGURE 10.5 Hydraulic design information sign should be installed on the main sprinkler system riser near the alarm check valve, dry pipe valve, pre-action valve, or deluge valve supplying the system.

to the water supply for the building as well as the piping and control valves. Note the static pressure on the supply side of the alarm check valve or dry pipe valve. Slowly open the main drain and flow water until clear, usually around two minutes. Compare the residual pressure to previous tests. Flow tests to determine the proper function of the flow switch hooked into the building fire alarm system are conducted with the use of the inspector's test valve. This valve is usually a part of the floor zone assembly. Dry sprinkler systems have an inspector's test outlet in addition to the inspector's test valve. This device is designed to simulate the flow of a single sprinkler head. The device is located at the most remote section of the dry sprinkler system. When the valve is opened, water must flow through the opening within one minute and an alarm must sound in five minutes or less. All testing of sprinkler systems, no matter what type or interval, should be documented. The AHJ should review this documentation during routine inspections once again to insure that the testing is being performed as required.

STANDPIPES MAINTENANCE AND TESTING

Following the construction of a new standpipe system, all components of the system, including the FDC, should be tested to ensure proper installation and periodically after that. New systems should be tested before occupancy of the building is allowed. Contractors that installed the system need to provide the appropriate contractor's material and test certificate to the AHJ before any testing of the system takes place. Samples of such certificates can be found in NFPA 14. All underground piping needs to be flushed according to NFPA 24, *Standard for the Installation of Private Fire Service Mains and Their Appurtenances.* Hose connection threads should be tested to verify that they are compatible with threads used by the local fire department. All new system piping needs to be hydrostatically tested at not less than 200 psi for two hours. Contractor certification should be provided to the AHJ for all hydrostat testing of the system piping. It is not necessary that the AHJ witness personally the hydrostat testing. The contractor can witness the tests and provide the documentation required. Water flow tests need to be conducted at the most remote hose connection (usually the roof connection) to make sure the system can deliver the required flow at the proper pressure. Main drain tests should also be conducted and static and residual pressures recorded on the contractor's test certificate. Automatic and semiautomatic dry systems should have a full function test conducted. When the hose valve is opened, water should flow from the valve at a minimum rate of 250 gpm (946 L/min) within three minutes. All valves should be fully opened and closed to ensure proper operation, including each fire department hose connection. Systems with fire pumps must be tested with the fire pump running. Fire department connections should also be flow tested to make sure water actually gets to the sprinkler and standpipe systems and there are no blockages in the piping system. Periodic maintenance needs to be provided for fire department connections according to the requirements of NFPA 25. Protective caps or plugs need to be in place to prevent obstructions from being placed inside. Locking caps can also be used. The female swivel connections should be free to spin. They should be lubricated with powdered graphite or other lubricant. Threads should be smooth and undamaged.

Testing of standpipes and other fire protection systems is governed by NFPA 25, *Inspection, Testing, and Maintenance of Water Based Fire Protection Systems*. Standpipes are required to be maintained per the frequencies outlined in the tables of NFPA 25. Testing and maintenance timetables range from weekly to every five years. Weekly maintenance involves only visual inspections. Alarm devices are tested quarterly with other fire alarm tests. Hoses and storage devices are visually inspected annually. Hose connections and valves are maintained annually. A main drain test is conducted annually. Hose nozzles and hose storage devices are tested annually. Hose pressure valves, hydrostatic tests, and flow tests are conducted every five years. Records of all inspections, maintenance, and testing should be maintained for review by the AHJ upon request. Private fire service mains are also required to be inspected, maintained, and tested periodically according to tables in NFPA 25. Tables identify the type of activity required—inspection, test, or maintenance—and the frequency—quarterly, annually, and every five years.

FIRE PUMP MAINTENANCE AND TESTING

Building fire pumps, like fire department engines, need to be inspected, operated, and tested periodically to ensure they are available when needed during an emergency. Inspection and test frequencies range from weekly to annually and are outlined in NFPA 25. Fire prevention bureaus as well as firefighters should know which buildings have fire pumps. They should make sure at least annually that the proper testing and maintenance is being provided according to the timetables outlined in NFPA 25. Fire pumps, when installed, require an acceptance test that includes a

FIGURE 10.6 This test is much like the original acceptance test and requires water to be flowed through the fire pump test header.

representative of the pump manufacturer to be present during the test. The manufacturer's representative is there to witness or conduct the test and make sure that the pump is installed according to manufacturer's instructions. It is highly recommended that the AHJ be present for the fire pump testing as well. Criteria for the acceptance testing of stationary fire pumps is outlined in NFPA 20, *Standard for Installation of Stationary Fire Pumps for Fire Protection*, and NFPA 25. Initial acceptance test reports should be produced and a copy given to the AHJ and should include a flow graph for the pump. This graph is important to compare future test graphs against to determine if there is any deterioration of the pumping capacity. Following the acceptance testing there are periodic tests required for all fire pumps. The pump should be run weekly without flowing water for a period of ten minutes for an electric driver and thirty minutes for a diesel driver. In addition to the functional test, there are various inspection requirements as well. Qualified operating personnel should be present during the weekly pump operation and annual fire pump test flow tests. Other testing is required annually to include a full flow fire pump test. This test is much like the original acceptance test and requires water to be flowed through the fire pump test header (figure 10.6). The AHJ does not need to be present; however, test forms including the flow graph need to be filled out by the person conducting the test and made available for inspection by the AHJ upon request.

PORTABLE FIRE EXTINGUISHERS MAINTENANCE AND TESTING

NFPA 10, *Standard for Portable Fire Extinguishers*, outlines the required testing and intervals for fire extinguisher maintenance. Extinguishers are required to be maintained on an annual basis with visual inspections monthly and hydrostatic testing of pressure vessels every five to twelve years, depending on the agent used in the extinguisher. Monthly inspections should include checking the nozzle for obstructions, making sure the gauge is in the green operating range, making sure the security tag is intact, checking the extinguisher for obvious damage, and initialling the record-keeping card attached to the extinguisher. Stored pressure extinguishers are required to be fully disassembled annually to receive complete maintenance. The extinguisher is discharged prior to disassembly to determine whether it is functioning properly. Conductivity tests are required for carbon dioxide extinguisher hose assemblies. Those that fail must be replaced. When fire extinguishers are removed from service for maintenance and testing, they must be replaced with a suitable extinguisher of equal or greater rating. Monthly visual inspections are required for all portable fire extinguishers and should include: location in designated place; no obstruction to access or visibility; operating instructions on nameplate legible and facing outward; safety seals and tamper indicators not broken or missing; fullness determined by weighing or hefting; examination for obvious physical damage, corrosion, leakage, or clogged nozzle; pressure gauge reading or indicator in the operable range or position; HMIS label in place. HMIS stands for Hazardous Materials Information System. This is a right-to-know system designed for people who work around chemicals in the workplace; in this case, fire extinguisher chemicals. Under the law, workers have a right to know what chemicals they are working around and what the hazards of those chemicals are.

Depending on the size of your facility, it can become very time consuming to complete the monthly visual inspections required by NFPA 10. Once every month, each fire extinguisher is required to be visually inspected and the inspection recorded with a date on a tag placed on each fire extinguisher. It's really up to the authority having jurisdiction whether they choose to enforce this requirement. This task can be accomplished by employees and you do not have to hire a fire protection company, which will likely save money, but it will still take time to do it. For the annual tests, it is necessary to contact a fire protection company to conduct the annual inspection and testing. Each extinguisher should be provided with a current inspection tag that indicates the month and year of the maintenance and identifies the person performing the service.

Every six years, extinguishers that require a twelve-year hydrostat test must be emptied and appropriate maintenance conducted. Every twelve years, the fire extinguisher container needs to undergo hydrostatic testing. This cycle usually applies to dry chemical and dry powder extinguishers. Carbon dioxide and pressurized water fire extinguishers are required to be hydrostat tested every five years. Fire extinguisher tanks are stamped with the date of the hydrostat test, so it is easy to determine when the last test occurred. NFPA 10 contains complete details of required procedures for the testing and maintenance of portable fire extinguishers and should be consulted to insure compliance with all details. Keep in mind that Class B extinguishers designed for commercial cooking operations need to be replaced with Class K agents during the six-year dumping or the hydrostatic test of the container, what ever comes first.

ELEVATORS MAINTENANCE AND TESTING

Local building code officials sometimes provide maintenance inspections of elevator operations. In some states they have state elevator inspection departments within other state agencies or local agencies that oversee the acceptance and periodic testing of elevators within the jurisdiction. In the state of Maryland, elevator inspections are conducted by the Division of Labor and Industry Elevator and Escalator Safety. Acceptance testing is important to ensure that the elevators have been installed according to code and are safe to operate. Building codes generally contain the regulations that apply to elevator construction. ASME Standard A17.1 is referenced by most codes for detailed requirements pertaining to elevator design and use. Elevators are also required by ASME A17.1 to have periodic maintenance. Section 1206 covers elevator maintenance requirements for elevators. Testing of passenger and freight elevators is covered in sections 1001–1003. Inspection and tests are required every six months as a minimum. The list of inspection and testing requirements is pretty extensive and persons interested in details should refer to ASME 17.1. Chapter 9 of NFPA 101, the *Life Safety Code*, requires the monthly operation of the fire service function of all elevators. A written record of such operations needs to be kept in the elevator machine room for the inspection of the authority having jurisdiction.

GENERATORS AND EMERGENCY LIGHTING UNITS MAINTENANCE AND TESTING

Emergency lighting systems should be continuously in operation or be capable of repeated automatic operation without manual intervention. Routine maintenance should be based upon the manufacturer's recommendations, instruction manuals, requirements of the authority having jurisdiction and NFPA code requirements. NFPA 101, the *Life Safety Code*, requires that a functional test be conducted on every emergency lighting system at thirty-day intervals for no less than thirty seconds. An annual full function test is required for battery-powered emergency lighting systems for no less than ninety minutes. The emergency lighting is required to be fully operational throughout the ninety-minute test. Self-testing or self-diagnostic emergency lighting systems and computer-based systems are permitted and must also follow the testing requirements listed above. Written records of visual inspections and functional tests should be maintained by the owner or representative and available to the AHJ upon request. Response personnel can ask to see such records if spot tests reveal units not working. Testing requirements for emergency generators are located in NFPA 110, *Standard for Emergency and Standby Power Systems*. Sample maintenance logs and maintenance schedules are located in Annex A. Generator operational tests are conducted at a minimum of once a month. Generator tests are required to be initiated at an automatic transfer switch and be on alternate power for a period of not less than thirty minutes. Transfer switches are required to be tested and maintained as well. Storage batteries and other generator components are required to be inspected weekly. Generator test records must also be maintained for the AHJ.

Whenever the generator is out of service, consideration should be given to temporarily utilizing a portable or alternative source. Test and maintenance records should be kept and made available to the AHJ upon request.

HVAC AND SMOKE CONTROL SYSTEMS MAINTENANCE AND TESTING

Maintenance and testing of HVAC systems, fire alarm devices, and dampers is covered in NFPA 90A and NFPA 72, the *National Fire Alarm Code*. Acceptance testing is very important to insure that smoke and fire dampers and duct detectors have been properly installed and will work under fire conditions. The University of Maryland, Baltimore, built a new eleven-story high-rise dental school and occupied the building in 2006. During construction, the fire dampers throughout the building were tested following installation. Testing found numerous dampers that failed to close all the way when the fusible link was melted. Consultation with the factory uncovered the fact that the dampers were improperly assembled at the factory, resulting in the damper failures in the field. This resulted in a nationwide recall of the dampers in question. Without testing fire and smoke dampers, there is no way to be certain that they are properly installed or will work under fire conditions. All detection devices that are associated with HVAC and smoke control need to be tested as well in accordance with NFPA 72, the *National Fire Code*, to make sure that they function as intended during fire alarm conditions. Smoke detectors should be tested with live smoke, and heat detectors, if any, should be tested with a heat gun or similar device. Smoke detectors should also be sensitivity tested according to the manufacturer's recommendations. Both manual and automatic HVAC controls should be operated to make sure they function properly.

References

NATIONAL FIRE PROTECTION AGENCY

NFPA 1. 2006. *Uniform fire code™*. Quincy, Mass.: National Fire Protection Association.

NFPA 10. 2002. *Standard for portable fire extinguishers*. Quincy, Mass.: National Fire Protection Association.

NFPA 11. 2005. *Standard for low-, medium-, and high-expansion foam*. Quincy, Mass.: National Fire Protection Association.

NFPA 12. 2005. *Standard on carbon dioxide extinguishing systems*. Quincy, Mass.: National Fire Protection Association.

NFPA 12A. 2004. *Standard on Halon 1301 fire extinguishing systems*. Quincy, Mass.: National Fire Protection Association.

NFPA 13. 2002. *Standard for the installation of sprinkler systems*. Quincy, Mass.: National Fire Protection Association.

NFPA 13D. 2002. *Standard for the installation of sprinkler systems in one- and two-family dwellings and manufactured homes*. Quincy, Mass.: National Fire Protection Association.

NFPA 13E. 2005. *Recommended practice for fire department operations in properties protected by sprinkler and standpipe systems*. Quincy, Mass.: National Fire Protection Association.

NFPA 13R. 2002. *Standard for the installation of sprinkler systems in residential occupancies up to and including four stories in height*. Quincy, Mass.: National Fire Protection Association.

NFPA 14. 2003. *Standard for the installation of standpipe and hose systems*. Quincy, Mass.: National Fire Protection Association.

NFPA 15. 2001. *Standard for water spray fixed systems for fire protection*. Quincy, Mass.: National Fire Protection Association.

NFPA 16. 2003. *Standard for the installation of foam-water sprinkler and foam-water spray systems*. Quincy, Mass.: National Fire Protection Association.

NFPA 17. 2002. *Standard for dry chemical extinguishing systems*. Quincy, Mass.: National Fire Protection Association.

NFPA 17A. 2002. *Standard for wet chemical extinguishing systems*. Quincy, Mass.: National Fire Protection Association.

NFPA 20. 2003. *Standard for the installation of stationary pumps for fire protection*. Quincy, Mass.: National Fire Protection Association.

NFPA 22. 2003. *Standard for water tanks for private fire protection*. Quincy, Mass.: National Fire Protection Association.

NFPA 24. 2002. *Standard for the installation of private fire service mains and their appurtenances*. Quincy, Mass.: National Fire Protection Association.

NFPA 25. 2002. *Standard for the inspection, testing, and maintenance of water-based fire protection systems*. Quincy, Mass.: National Fire Protection Association.

NFPA 30. 2003. *Flammable and combustible liquids code*. Quincy, Mass.: National Fire Protection Association.

NFPA 37. 2002. *Standard for the installation and use of stationary combustion engines and gas turbines*. Quincy, Mass.: National Fire Protection Association.

NFPA 54. 2006. *National fuel gas code*. Quincy, Mass.: National Fire Protection Association.

NFPA 58. 2004. *Liquefied petroleum gas code*. Quincy, Mass.: National Fire Protection Association.

NFPA 70. 2005. *National electrical code®*. Quincy, Mass.: National Fire Protection Association.

NFPA 72. 2002. *National fire alarm code®*. Quincy, Mass.: National Fire Protection Association.

NFPA 90A. 1999. *Standard for the installation of air-conditioning and ventilating systems*. Quincy, Mass.: National Fire Protection Association.

NFPA 90B. 2006. *Standard for the installation of warm air heating and air-conditioning systems*. Quincy, Mass.: National Fire Protection Association.

NFPA 96. 2004. *Standard for ventilation control and fire protection of commercial cooking operations*. Quincy, Mass.: National Fire Protection Association.

NFPA 99. 2005. *Standard for health care facilities*. Quincy, Mass.: National Fire Protection Association.

NFPA 101. 2006. *Life safety code®*. Quincy, Mass.: National Fire Protection Association.

NFPA 110. 2005. *Standard for emergency and standby power systems*. Quincy, Mass.: National Fire Protection Association.

NFPA 111. 2005. *Standard on stored electrical energy emergency and standby power systems*. Quincy, Mass.: National Fire Protection Association.

NFPA 170. 2006. *Standard for fire safety and emergency symbols*. Quincy, Mass.: National Fire Protection Association.

NFPA 241. 2004. *Standard for safeguarding construction, alteration, and demolition operations*. Quincy, Mass.: National Fire Protection Association.

NFPA 291. 2002. *Recommended practice for fire flow testing and marking of hydrants*. Quincy, Mass.: National Fire Protection Association.

NFPA 704. 2001. *Standard system for the identification of the hazards of materials for emergency response*. Quincy, Mass.: National Fire Protection Association.

NFPA 720. 2005. *Standard for the installation of carbon monoxide (CO) warning equipment in dwelling units*. Quincy, Mass.: National Fire Protection Association.

NFPA 750. 2003. *Standard on water mist fire protection systems*. Quincy, Mass.: National Fire Protection Association.

NFPA 1141. 1998. *Standard for fire protection in planned building groups*. Quincy, Mass.: National Fire Protection Association.

NFPA 1142. 2001. *Standard on water supplies for suburban and rural fire fighting*. Quincy, Mass.: National Fire Protection Association.

NFPA 1963. 2003. *Standard for fire hose connections*. Quincy, Mass.: National Fire Protection Association.

NFPA 2001. 2004. *Standard on clean agent fire extinguishing systems*. Quincy, Mass.: National Fire Protection Association.

NFPA 5000. 2003. *Building construction and safety code®*. Quincy, Mass.: National Fire Protection Association.

OTHER REFERENCES

American Society of Mechanical Engineers. 1999. ASME A17.1. *Safety code for elevators and escalators: An American National Standard*, New York: Author.

American Society of Mechanical Engineers. 1999. ASME A17.4. *Guide for emergency personnel*. New York: Author.

Blossom, David R. 2005. Fire Pumps: The invisible firefighter. *Fire Engineering*. Retrieved August, 2007, from http://www.fireengineering.com/articles/article_display.html?id=235068.

Cote, Arthur E., ed. 2003. *NFPA fire protection handbook.* 19th ed. Quincy, MA: National Fire Protection Association.

International Code Council. 2003. *International building code.* Country Club Hills, IL: Author.

International Code Council. 2003. *International fire code.* Country Club Hills, IL: Author.

McGrail, David M. 2005. Fire company standpipe operations: Pressure-regulating devices. *Fire Engineering,* February.

Routley, J. Gordon, Jennings, Charles, and Mark Chubb. 1991. *High-rise office building fire, One Meridian Plaza, Philadelphia, Pennsylvania.* Emmitsburg, MD: United States Fire Administration (USFA) Technical Report Series.

Texas Women's University, Environmental Health and Safety. (nd). *Emergency elevator rescue.* Retrieved August, 2007, from http://www.twu.edu/rm/ehs/elevator.htm.

United States Fire Administration, National Fire Academy. (nd). *Coffee break training.* Retrieved August, 2007, from http://www.usfa.dhs.gov/nfa/coffee-break/.

University of Michigan. (nd). *Elevator shop emergencies, What do I do?* Retrieved August, 2007, from http://www.plantops.umich.edu/maintenance/shops/Elevator/Emergencies.html.

Index

Printed and bound by CPI Group (UK) Ltd, Croydon, CR0 4YY

24/10/2024

01778277-0010